P9-CDV-464

ROBOTS

A NEW AGE OF BIONICS, DRONES & ARTIFICIAL INTELLIGENCE

HEARST BOOKS
New York

An Imprint of Sterling Publishing
1166 Avenue of the Americas
New York, NY 10036

Every effort has been made to ensure that all the information in this book is accurate. However, due to differing conditions, tools, and individual skills, the publisher cannot be responsible for any injuries, losses, and/or other damages that may result from the use of the information in this book.

ISBN 978-1-61837-168-3

Distributed in Canada by Sterling Publishing
c/o Canadian Manda Group, 664 Annette Street
Toronto, Ontario, Canada M6S 2C8
Distributed in the United Kingdom by GMC Distribution Services
Castle Place, 166 High Street, Lewes, East Sussex, England BN7 1XU
Distributed in Australia by Capricorn Link (Australia) Pty. Ltd.
P.O. Box 704, Windsor, NSW 2756, Australia

For information about custom editions, special sales, and premium and corporate purchases, please contact Sterling Special Sales at 800-805-5489 or specialsales@sterlingpublishing.com.

Manufactured in China

2 4 6 8 10 9 7 5 3 1

www.sterlingpublishing.com

Popular Mechanics

ROBOTS

A NEW AGE OF BIONICS, DRONES & ARTIFICIAL INTELLIGENCE

Edited by Daniel H. Wilson

HEARST BOOKS
New York

Google
OSHKOSH
Carnegie Mellon
NREC

CONTENTS

INTRODUCTION

Witnessing the rise of the robots is a wondrous experience—and a frightening one.

N THE LAST CENTURY, ROBOTS IN SCIENCE-FICTION HAVE RISEN UP AGAINST their human masters; visited Earth from distant stars to pass judgment on our civilization; and relentlessly chased our leaders back in time in an effort to eradicate humanity itself.

Robots have always lurked in our stories as bogeymen, and now their real-world counterparts are steadily creeping into our lives. And to be perfectly honest, I find myself deeply excited.

Since 1956, when a Stanford professor named John McCarthy coined the term "artificial intelligence," our thinking machines have evolved from playing chess on a board with sixty-four squares to playing *Jeopardy!* with open-ended questions asked out loud. This vastly increased brainpower has been accompanied by equivalent physical progress in manipulation, locomotion, and dexterity.

We are reaching an unprecedented moment in history, in which robots have evolved from things into partners and allies with the ability to make their own decisions. Unlike every other tool ever built, robots are not simply objects to be manipulated, but agents that can think and act alongside us in the world.

Our latest smart machines soar through hostile skies as warfighters, comb through online data for commerce, and in some cases, even wriggle through our veins to provide medical treatment. From the Aeolis Palus plains on Mars to the sprawling corridors of darkened warehouses, a flood of advanced technology is rapidly growing to solve a dazzling variety of hard problems.

While earning a doctorate degree in robotics from Carnegie Mellon University (CMU), I witnessed the development of a vision system that could simultaneously track every individual ant in an entire colony. Disembodied robot legs were a common sight, twisting and writhing as they taught themselves novel modes of locomotion. And proof-of-concept mobile robots often wheeled around the CMU Robotics Institute, usually missing a panel or two, and occasionally shoving their way into an elevator or in line for coffee.

I am excited today because the basic research I witnessed ten years ago has finally escaped the laboratory and begun to have an effect on the rest of the world. This is why I chose to organize this book into six sections, based on the six areas in which robots are making the biggest impact: in our daily lives, our cities, the skies, the military, medicine, and outer space.

It is my pleasure to introduce you to the next wave of machines. In this book, you'll discover robots designed to cooperate with us, speak to us, and recognize our faces, gestures, and emotions; as well as robots who are joining us in the form of self-driving cars, artificially intelligent personal assistants, and even embodied in the architecture of our homes.

As we progress deeper into the Information Age, new and astonishing robots will continue to proliferate in our homes, cities, and places of work. How incredibly lucky are we to live during this amazing time? Together, we have the opportunity to learn more about our robotic partners and embrace an awe-inspiring future.

Although we may feel both fear and wonder, remember that this is not the end of days–it's the beginning of a wonderful new world.

—Daniel H. Wilson

SARCOS
Bicycles
Park City
Veterans
welcoming place
WHITE

Chapter

01

ROBOTS IN DAILY LIFE

CAN ROBOTS BE TRUSTED?

Humans have feared a robotic uprising since the machines first appeared in science fiction. Today, experts caution against a more insidious threat: We might like living with them too much.

BY **ERIK SOFGE**

This article was published in *Popular Mechanics* in 2010.

EING HACKED BY A ROBOT REQUIRES much less hardware than I expected. There's no need for virtual-reality goggles or 3D holograms. There are no skullcaps studded with electrodes, no bulky cables or hair-thin nanowires snaking into my brain. Here's what it takes: one pair of alert, blinking eyeballs.

I'm in the Media Lab, part of MIT's sprawling campus in Cambridge, Mass. Like most designated research areas, the one belonging to the Personal Robots Group looks more like a teenage boy's bedroom than some pristine laboratory—it bursts with knotted cables, old pizza boxes, and what are either dissected toys or autopsied robots. Amid the clutter, a 5-foot-tall, three-wheeled humanoid robot boots up and starts looking around the room. It's really looking, the oversize blue eyes tracking first, and the white, swollen, doll-like head following, moving and stopping as though focusing on each researcher's face. Nexi turns and looks at me. The eyes blink. I stop talking, midsentence, and look back. It's as instinctive as meeting a newborn's roving eyes. *What do you want?* I feel like asking. *What do you need?* If I was hoping for dispassionate, journalistic distance—and I was—I never had a chance.

"Right now it's doing a really basic look around," researcher Matt Berlin says. "I think it's happy, because it has a face to look at." In another kind of robotics lab, a humanoid bot might be motivated by a specific physical goal— cross the room without falling, find the appropriate colored ball and give it a swift little kick. Nexi's functionality is more ineffable. This is a social robot. Its sole purpose is to interact with people. Its mission is to be accepted.

That's a mission any truly self-aware robot would probably turn down. To gain widespread acceptance could mean fighting decades of robot-related fear and loathing. Such stigmas range from doomsday predictions of machines that inevitably wage war on mankind to the belief that humanoid robots will always be hopelessly unnerving and unsuitable companions.

For Nexi, arguably the biggest star of the human-robot interaction (HRI) research field, fame is already synonymous with fear. Before visiting the

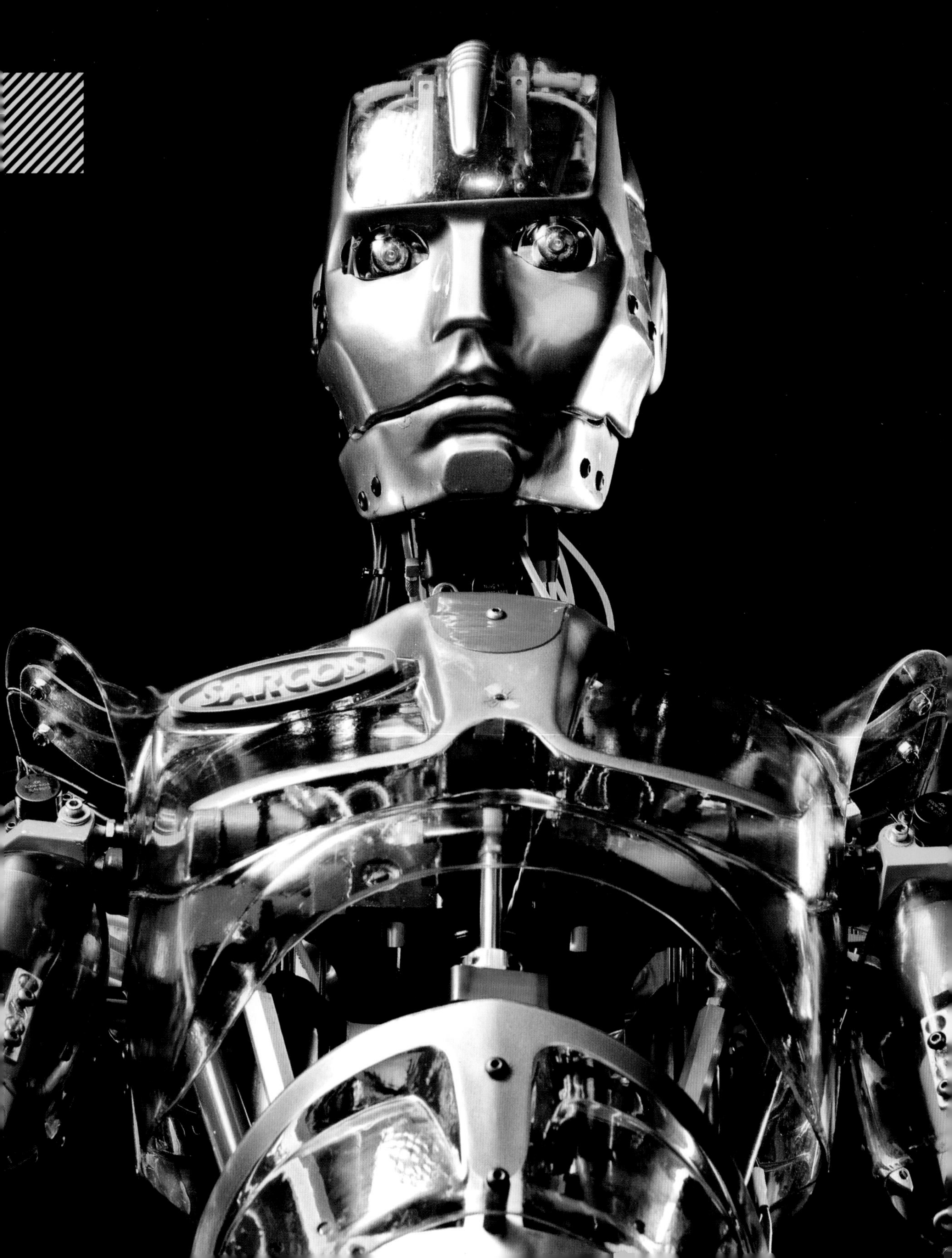
SARCOS

Media Lab, I watched a video of Nexi that's been seen by thousands of people on YouTube. Nexi rolls into view, pivots stiffly to face the camera, and introduces itself in a perfectly pleasant female voice. If the goal was to make Nexi endearing, the clip is a disaster. The eyes are big and expressive, the face is childish and cute, but everything is just slightly off, like a possessed doll masquerading as a giant toddler. Or, for the existentially minded, something more deeply disturbing—a robot with real emotions, equally capable of loving and despising you. Viewers dubbed its performance "creepy."

Now, staring back at Nexi, I'm an instant robot apologist. I want to shower those clips with embarrassingly positive comments, to tell the haters and the doubters that the future of HRI is bright. There's no way seniors will reject the meds handed to them by chattering, winking live-in-nurse bots. Children, no doubt, will love day-care robots, even if the bots sometimes fail to console them, or grind to an unresponsive halt because of buggy software or faulty battery packs. To turn today's faceless Roombas into tomorrow's active, autonomous machine companions, social robots need only to follow Nexi's example, tapping into powerful, even uncontrollable human instincts.

That's why Nexi's metallic arms and hands are drifting around in small, lifelike movements. It's why Nexi searches for faces and seems to look you in the eye. When it blinks again, with a little motorized buzz, I realize I'm smiling at this thing. I'm responding to it as one social, living creature to another. Nexi hasn't said a word, and I already want to be its friend.

As it turns out, knowing your brain is being hacked by a robot doesn't make it any easier to resist. And perhaps that's the real danger of social robots. While humans have been busy hypothesizing about malevolent computers and the limits of rubber flesh, roboticists may have stumbled onto a more genuine threat. When face to face with actual robots, people may become too attached. And like human relationships, those attachments can be fraught with pitfalls: How will grandma feel, for example, when her companion bot is packed off for an upgrade and comes back a complete stranger?

When a machine can push our Darwinian buttons so easily, dismissing our deep-seated reservations with a well-timed flutter of its artificial eyelids, maybe fear isn't such a stupid reaction after all. Maybe we've just been afraid of the wrong thing.

OBOTS BEGAN SCARING US LONG BEFORE they existed. In 1921, the Czech play *R.U.R.*, or *Rossum's Universal Robots*, simultaneously introduced the word "robot" and the threat of a robot apocalypse. In a proclamation issued in the play's first act, the robots, built as cheap, disposable laborers, make their intentions clear: "Robots of the world, we enjoin you to exterminate mankind. Don't spare the men. Don't spare the women." The origins of the evil robot can be traced back even further, but *R.U.R.*'s new species of bogeyman was all the rage in the pulp sci-fi of the '40s and '50s—well before the actual research field of robotics. In fact, *I, Robot* author Isaac Asimov coined the term "robotics" at the same time that he began developing ethical laws for robots in his short stories.

By the time Arnold Schwarzenegger's T-800 gunned down an entire police precinct in the 1984 movie *The Terminator*, the robot insurgency had become one of pop culture's most entrenched clichés. The film has since become shorthand for a specific fear: that artificial intelligence (AI) will become too intelligent, too obsessed with self-preservation. *The Terminator* colors the way we think about robots, AI, and even the booming business of unmanned warfare. The Office of Naval Research, among others, has studied whether ethical guidelines will be needed for military robots, and in a 2008 preliminary report the authors tackle the bleakest possible endgame: "*[The] Terminator* scenarios where machines turn against us lesser humans."

But according to Patrick Lin, an assistant professor of philosophy at California Polytechnic State University and an ethics fellow at the U.S. Naval Academy, the need for ethical bots isn't restricted to the battlefield. "Social robots probably pose a greater risk to the average person than a military robot," Lin says. "They won't be armed, but we will be coming face to face with them quite soon."

That, of course, is precisely the kind of quote reporters work hard to publish. The media homes in on juicy details about the hypothetical danger of self-organizing AI, and the prospect of amoral robots gunning down civilians. But the real threats posed by robots may have nothing to do with *The Terminator* scenario. Because compared to even the dumbest armed insurgent, robots are practically brain-dead.

IT'S NOT THE HARDWARE THAT MAKES THE EVIL ROBOT ONE OF WESTERN CULTURE'S MOST POWERFUL MYTHS. IT'S THE SOFTWARE, THE ARTIFICIAL INTELLIGENCE (AI) THAT TURNS MACHINES INTO MONSTERS. HERE ARE THE MOST ICONIC EXAMPLES OF MALEVOLENT AI WITH THE FEARS EACH INSPIRED.

Golem of Prague

Taught us to fear: unstable artificial intelligence

In folk tales, the Golem of Prague was sculpted from river mud and animated with magic, but its design is undeniably robotic–big and emotionless. Its limited AI is also familiar: The Golem floods a house when no one tells it to stop fetching water. In later versions of the myth, it loses its mind.

Frankenstein's Monster

Taught us to fear: artificial genius

Critics call Frankenstein the world's most influential evil-robot story. The monster's whip-smart mind is his own undoing. He learns to speak and read in months and to resent his creator just as quickly. Frankenstein refuses to build a mate, fearing a superior, malevolent race that would destroy mankind.

Radius

Taught us to fear: organized robotic insurrection

Like Frankenstein's monster, the robots in the play *R.U.R.* are flesh-and-blood murderers. These robots are produced from factory-grown organs, and successfully wipe out the human race. The robot leader, Radius, doesn't mince words, saying, "I wish to be the master of people."

The Machines

Taught us to fear: a less deadly but more secret insurrection

In his short story "The Evitable Conflict," Isaac Asimov granted the machines control of the world economy. And they prove overzealous. Hollywood supplied the melodrama, turning AI's quiet financial coup into the mass house-arrest of mankind in the movie *I, Robot*.

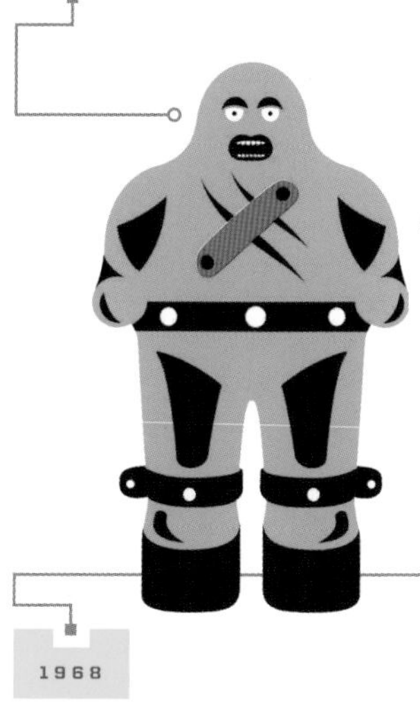

HAL 9000

Taught us to fear: AI-controlled systems

The singsong condescension in HAL 9000's voice should have been a warning sign. But by the time *2001: A Space Odyssey* leaps from sci-fi to horror, it's too late–the AI jettisons the human crew members it considers to be a liability to the spacecraft's mission. Like Asimov's machines, HAL isn't malicious, just a little too smart for our own good.

T-800/Skynet

Taught us to fear: networked, self-organizing AI

Skynet never appears on camera in *The Terminator*, but the movie's eponymous enforcer bears its message: The planet's not big enough for biological and artificial intelligence. We also don't see the advanced defense computer becoming self-aware. Instead, the movie shows the smoldering aftermath of war, giving an old myth its most powerful update.

ED-209

Taught us to fear: armed, autonomous robots

In *RoboCop*, ED-209 has the cognitive powers of a very smart police dog and the firepower of an attack chopper. And, like dogs trained for violence, ED-209 sometimes bites the wrong person: In one of the movie's most memorable scenes, the security bot botches its own sales demo by gunning down an unarmed civilian.

The Machines

Taught us to fear: everything in *The Terminator*, and robot slavers

The machines of *The Matrix* are a deliriously twisted race of AIs. They turn prisoners into battery packs, craft vast virtual worlds to keep us occupied and, as evidenced by Agent Smith, are capable of abject hatred. The real horror of *The Matrix* (sequels aside) is the prospect of machines not only conquering mankind, but toying with our defeated species.

Take Nexi, for example. Considered to be one of the most advanced social robots in the world, Nexi can understand only the most basic vocal instructions. During my visit, it couldn't even do that–it was in the process of being loaded with behavioral software developed for another MIT robot, the fuzzy, bigeared Leonardo. Now in semi-retirement–its motors have gone rickety–Leonardo learns from humans such lessons as which blocks fit into a given puzzle, or which stuffed animal is "good" and which it should be afraid of. The implications are of the mind-blowing variety: a robot that listens to what we say and learns to crave or fear what we tell it to. Programmed with Leonardo's smarts, "maybe in a year Nexi will be able to have a conversation with you that's very boring," MIT's Berlin says. "But it may be pretty interesting if you're trying to escape a burning building."

If David Hanson, the founder of Hanson Robotics, has his way, the Texas-based company's latest social robot, Zeno, could be talking circles around Nexi by the end of this year. At $2500, the 23-inch-tall humanoid robot would be a bargain, not because of its hardware but because of the code crammed into its cartoonish head. "The intelligent software can be aware of multiple people in a room," Hanson says. "It builds a mental model of who you are, what you like, and what you said. We're getting to the point where it can hold an open-ended, open domain conversation." Hanson plans to roll out a cheaper mass market version in 2011 or 2012, with the same facial- and vocal recognition capabilities. His goal is to provide a powerful testbed for researchers, while also harnessing AI algorithms to make a robot toy that's actually fun for more than 15 minutes.

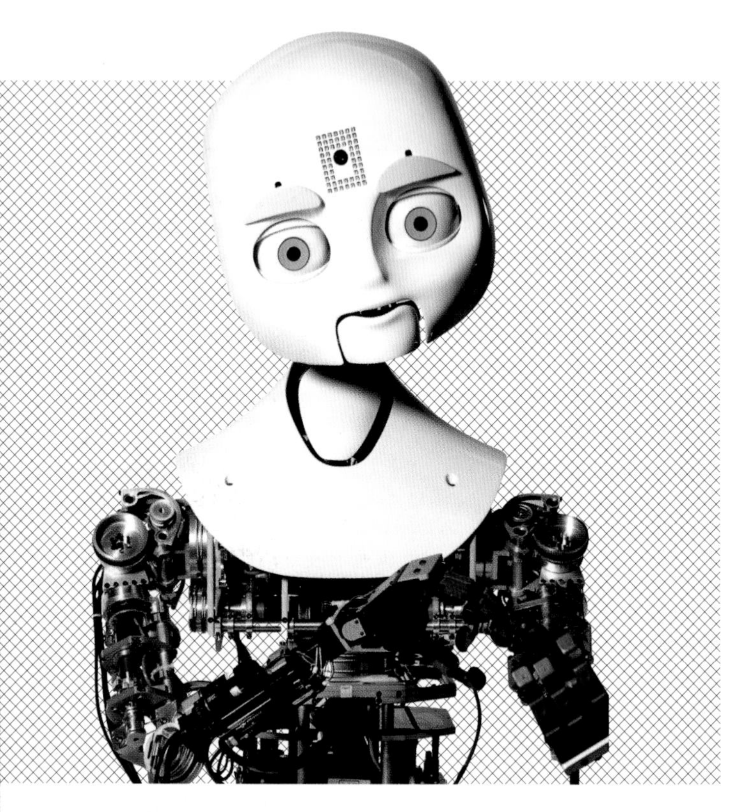

NEXI: *THE FIRST IN A PROPOSED CLASS OF MDS, OR MOBILE DEXTROUS SOCIAL, ROBOTS, NEXI IS THE MOST HIGH-PROFILE PROJECT IN MIT'S PERSONAL ROBOTS GROUP. IT IS PART OF AN APPROACH, PIONEERED AT MIT, CALLED EMBODIED AI–ARTIFICIAL INTELLIGENCE THAT, LIKE HUMAN INTELLIGENCE, IS TIED TO THE WORKINGS AND LIMITATIONS OF ITS OWN BODY.*

But for all of Nexi's and Zeno's social skills and painstaking simulation of emotional life, the bots are creatures of instinct, not introspection. Tracking software finds the human who's speaking, a keyword triggers a scripted response, and when you leave the room, they don't imagine where you've gone, whether the conversation helped or hurt you, or how to overthrow your government. "It's very difficult for an artificial intelligence to project in a physical sense," says Kevin Warwick, a professor of cybernetics at the University of Reading in England. "A robot can think about eventualities, but it can't think even one step ahead about the consequences of its decisions."

There are, of course, researchers who foresee rapid progress in computational neuroscience leading to inevitable "strong AI," or artificial intelligence that's not simply finishing your sentence in a Google search box, but mimicking human thought. IBM's Blue Brain Project, for one, is energizing doomsayers with its goal of creating a virtual brain, potentially as soon as 2019. Still, without a neurological map of our own sense of consequence or morality, the breakthroughs that would allow for a truly power-hungry or evil robot are nowhere in sight. Contemplating them is a little like debating the ethical pitfalls of unregulated teleportation. Until someone builds the Enterprise, why worry if Scotty is going to drunk-dial himself into your house?

Robots will not rise up en masse anytime soon. Nexi won't be e-mailing Zeno the "exterminate all humans" flier from *R.U.R.* to distribute among the world's Roombas, Predators, and assembly-

CAPABLE *OF GESTURING AND SPEAKING, SARCOS WAS BUILT IN 1997 TO TALK ABOUT TECHNOLOGY WITH CHILDREN AT THE CARNEGIE SCIENCE CENTER IN PITTSBURGH. SOCIAL ROBOTS LIKE IT ARE EXPECTED TO BE A $15 BILLION INDUSTRY BY 2015.*

line welding machines. It's a fantasy, or, at best, a debate for another century. And like many robot fears, it threatens to drown out a more rational debate, one that stems from the fact that robots fall through nearly every legal and ethical crack. "If an autistic patient charges a robot and tries to damage it, how should the robot respond?" asks Lin, who is also planning to develop ethical guidelines for social healthcare bots. "Should it shut down? It's an expensive piece of equipment–should it push back?" When the robots arrive in force, are we prepared for the collateral damage, both physical and psychological, they could inflict?

HEN OUR EYES SEE A ROBOT, one that we think is autonomous–moving, acting, functioning under its own power–our mirror neurons fire. These same neurons activate when we watch another animal move, and neuroscientists suspect they're associated with learning, by way of imitation. Mirror neurons could care less about a wax statue, or a remote-control drone. It's the autonomous robot that lights the fuse, tricking the mind into treating a mechanical device as a living thing.

And yet, like many aspects of human-robot interaction, the full repercussions are unknown. Science-fiction writers may have spent a half-century theorizing about the long-term effects of living with robots, but science is only getting started. While the field of HRI collects data and sorts out its methodologies, drawing solid conclusions can be impossible, or at least irresponsible. Take those mirror neurons, for example. Neuroscientists can watch them flip on, but the exact purpose of those neurons is up for debate.

Another, more common example of the brain's mysterious response to robots is often referred to as the uncanny valley–a poetic way of saying, "robots are creepy." Proposed in a 1970 paper by roboticist Masahiro Mori, the uncanny valley describes a graph showing that humans feel more familiar with, and possibly more comfortable toward, humanoid machines. Until, that is, the machine becomes too human-like, tripping the same psychological alarms associated with seeing a dead or unhealthy human. At that point, the graph collapses, and then rises again with the response to a real human being, or theoretically, a perfect android. Whether this is a distortion of our fight-or-flight instincts or something more complex, Mori's word choice was important—the uncanny is not naked fear, but a mix of familiarity and fear, attraction and repulsion. It's a moment of cognitive dissonance that the brain can't reconcile, like encountering a talking Christmas tree or a laughing corpse.

By academic standards, it's evocative, exciting stuff, describing what appears to be a widespread phenomenon. Nexi's unnerving YouTube clips seem like textbook examples, and the robot has plenty of unsettling company. The Japanese social bot CB2 (Child-robot with Biomimetic Body), with its realistic eyes, child-like proportions, and gray skin, evokes near-universal horror among bloggers and reporters. Another Japanese robot, KOBIAN, features a wildly expressive face, with prominent eyebrows and a set of fully formed, ruby-red lips. It, too, was instantly branded creepy by the Western press. The designers of those social bots were actually trying to avoid the uncanny—Asian labs are packed with photorealistic androids that leap headlong into the twitching, undead depths of Mori's valley.

But just as *The Terminator* scenario withers under scrutiny, the uncanny valley theory is nowhere near as tidy as it sounds. Based on those YouTube clips, I had expected my meeting with Nexi to be hair-curling. Instead, I can see my grin scattered across computer monitors in the Media Lab. Nexi's forehead-mounted, depth-sensing infrared camera shows my face as a black and gray blur, and the camera in its right eye portrays me in color. I watch as I slip from the monitors, Nexi's head and eyes smoothly tracking to the next face. I am not creeped out—I'm a little jealous. I want Nexi to look at me again.

"There are some very practical things that we do to make our robots not creepy," Berlin says. The secret to Nexi's success, apparently, is within arm's reach of the robot: a slightly battered hardcover book titled *The Illusion of Life: Disney Animation*—required reading for the Personal Robots Group. "We're making an animation, in real time," Berlin says. Like many animated characters, Nexi's features and movements are those of exaggerated humanity. When it reaches for an object, its arm doesn't shoot forward with eerie precision. It wastes time and resources, orienting its eyes, head and body, and lazily arcing its hand toward the target. Nexi is physically inefficient, but socially proficient.

How proficient? In interactions with hundreds of human subjects, including residents of three Boston-area senior centers, researchers claim that no one has run screaming from Nexi. Quite the opposite: Many seniors tried to shake the robot's hand, or hug it. At least one of them planted a kiss on it. "It interacts with people in this very social way, so people treat it as a social entity in an interpersonal way, rather than a machinelike way," Cynthia Breazeal, director of the Personal Robots Group, says. "In studies with Nexi, we've shown that if you have the robot behave and move in ways that are known to enhance trust and engagement, the reaction is the same as it is with people. You're pushing the same buttons."

That principle has proven true for CB2 and KOBIAN as well. The research leaders of both projects claim that the apprehension directed at their robots online and in the media never materializes in person. With the exception of one Thai princess, everyone who encountered CB2 liked it, according to Osaka University's Minoru Asada. A Japanese newspaper brought a group of elderly to visit KOBIAN. They were "deeply pleased and moved," Atsuo Takanishi, a professor of mechanical engineering at Waseda University, says, "as if the robot really had emotion."

VEN IF THE UNCANNY VALLEY ends up being more of a shallow trench, one that's easily leveled by actually meeting an android, the success of Nexi and company only raises a more profound question: Why do we fall so hard for robots?

"It turns out that we're vulnerable to attaching, emotionally, to objects. We are extremely cheap dates," says Sherry Turkle, director of the MIT Initiative on Technology and Self. "Do we really want to exploit that?" Turkle has studied the powerful bond that can form between humans and robots such as Paro, an almost painfully cute Japanese baby-seal-shaped therapy bot that squirms in your arms, coos when caressed and recharges by sucking on a cabled pacifier. She has also documented assumptions of intelligence and even emotion reported by children playing with robotic dolls. The effect that Paro, little more than an animatronic stuffed animal, had on senior citizens only reinforced her concerns. "Tell me again why I need a robot baby sitter?" Turkle asks. "What are we saying to the child? What are we saying to the older person? That we're too busy with e-mail to care for those in need?"

To researchers like Turkle, the widespread deployment of social robots is as risky as it is inevitable. With some analysts estimating a $15 billion market for personal robots by 2015, the demand for expressive machines is expected to be voracious. At the heart of Turkle's argument—a call for caution, essentially—is the fear of outsourcing human interaction to autonomous machines. Even more alarming are the potential beneficiaries of robotic companionship, from children in understaffed schools to seniors suffering from Alzheimer's. Enlisting an army of robots to monitor the young and the elderly could be a bargain compared to the cost of hiring thousands of teachers and live-in nurses. But how will the first generation to grow up with robotic authority figures and friends handle unpredictable human relationships? Without more data, a well-intended response to manpower shortage could take on the ethical and legal dimensions of distributing a new and untested antidepressant.

One possible solution is to scale back the autonomy and use social bots as puppets. Huggable, another robot from MIT's Personal Robots Group, is a teddy bear whose movements can be controlled through a Web browser. The researchers plan to use it to comfort hospitalized children; family members or doctors would operate it remotely. When I see Huggable, it's actually a teddy bear skeleton. The furry coat, which will eventually be replaced with one that includes pressure and touch-sensitive sensors, sits in a heap next to the bot as it fidgets. An open laptop shows the operator's view through Huggable's camera and a menu of simple commands, such as raising and lowering its arms, or aiming its head at my face.

CYNTHIA ***BREAZEAL, THE DIRECTOR OF MIT'S PERSONAL ROBOTS GROUP, SAYS SOCIAL ROBOTS REQUIRE SYSTEMS THAT ARE "SAVVY AND INTELLIGENT IN THEIR INTERACTIONS WITH PEOPLE," NOT SIMPLY COMPATIBLE WITH OBJECTS.***

For now, Huggable has no identity of its own. It's a high-tech ventriloquist's dummy channeling the voice of its operator, not a full-fledged social creature. In a recent paper describing the dangers of "parent" modes in Japanese robotic toys and the temptation to use robots as nannies, Noel Sharkey, a professor of artificial intelligence and robotics at the University of Sheffield in England, cited Huggable's lack of autonomy as a selling point. "Such robots do not give rise to the same ethical concerns as exclusive or near-exclusive care by autonomous robots," he wrote with a co-author. Semiautonomy might not cut payrolls, but it could be a safer way to roll out the first wave of social bots.

Sharkey's and Turkle's ominous point of view overlaps uncomfortably with the climate of fear that has always surrounded robots. And yet, nearly every researcher I spoke with agreed on a single point: We need ethical guidelines for robots, and we need them now. Not because robots lack a moral compass, but because their creators are operating in an ethical and legal vacuum. "When a bridge falls down, we have a rough-and-ready set of guidelines for apportioning out accountability," says P.W. Singer, a senior fellow at the Brookings Institution and author of *Wired for War*. "Now we have the equivalent of a bridge that can get up and move and operate in the world, and we don't have a way of figuring out who's responsible for it when it falls down."

In a debate steeped in speculation and short on empirical data, a set of smart ethical guidelines could act as an insurance policy. "My concern is not about the immediate yuck factor: What if this robot goes wrong?" says Chris Elliott, a systems engineer and trial lawyer who contributed to a recent Royal Academy report on autonomous systems. "It's that people will go wrong." Even if the large-scale psychological impact of social robots turns out to be zero, Elliott worries that a single mishap, and the corresponding backlash could reverse years of progress. Imagine the media coverage of the first patient killed by a robotic surgeon, an autonomous car that T-bones a school bus or a video clip of a robotic orderly wrestling with a dementia patient. "The law is way behind. We could reach a point where we're afraid to deploy new beneficial robots because of the legal uncertainty," Elliott says.

The exact nature of those guidelines is still anyone's guess. One option would be to restrict the use of each robotic class or model to a specific mission—nurse bots that can visit with patients within a certain age range, or elder-care bots that watch for dangerous falls but aren't built for small talk and snuggling. In the long run, David Hanson believes AI should be explicitly programmed to cooperate with humans, so that when robots self-evolve they have what he calls the "wisdom"

KOBIAN: *THIS JAPANESE INVENTION IS THE INTELLECTUAL LOVE CHILD OF A PAIR OF EARLIER WASEDA UNIVERSITY ROBOTS—ONE AN EXPRESSIVE HEAD, THE OTHER A HUMANOID BODY. THE RESULT IS WHAT ITS CREATORS CALL "AN EMOTIONAL HUMANOID" ABLE TO EXPRESS EMOTIONS WITH ITS ENTIRE BODY, POTENTIALLY ALLOWING PERSONAL ROBOTS TO BETTER COMMUNICATE WITH HUMANS.*

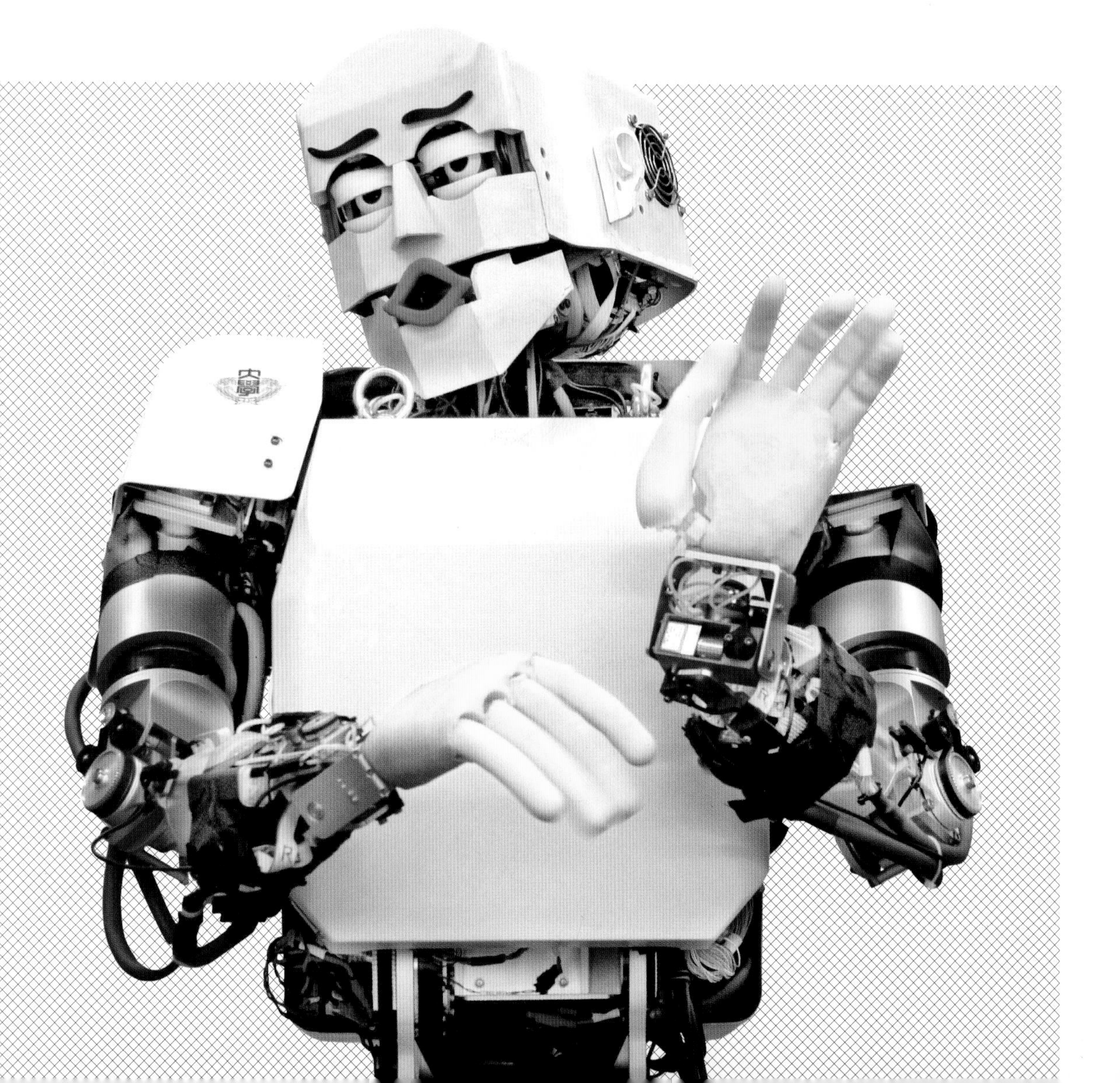

not to harm us. Cynthia Breazeal's take is more hard-nosed. "Now is certainly the time to start hammering things out," she says. "People should have a serious dialogue before these robots are in contact with vulnerable populations."

Philosophers, ethicists, lawyers, and roboticists have only begun the hard work of fleshing out Asimov's early code of robo-ethics. In the meantime, if there's a way to dismantle our long-standing, irrational fear of robots and head off any risk of a Luddite backlash, it might be up to robots such as Nexi.

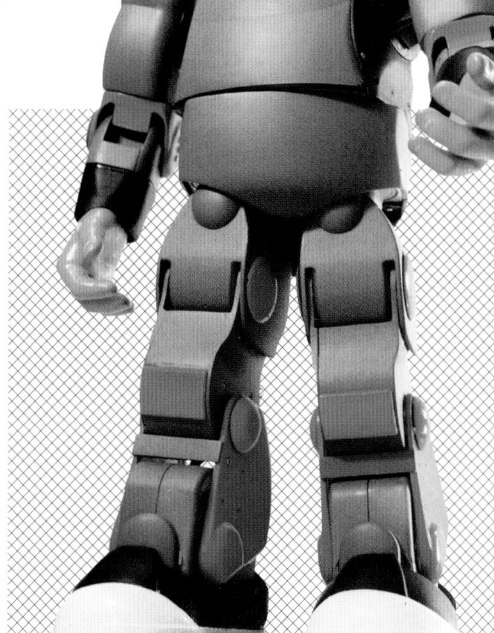

ZENO *IS MORE OF A BUSINESS PLAN THAN A STAND-ALONE HUMANOID, AN ATTEMPT BY HANSON ROBOTICS TO CHANNEL THE COMPANY'S BREAKTHROUGHS IN ARTIFICIAL SKIN AND SOCIAL LEARNING ALGORITHMS INTO A HYBRID ROBOT TOY AND DIRT-CHEAP RESEARCH TESTBED. IF ZENO CATCHES ON WITH KIDS, IT COULD BE THE WORLD'S BIGGEST—AND LEAST CONTROLLED—EXPERIMENT IN HUMAN-ROBOT INTERACTION.*

While I'm eyeing the gears and servos along Nexi's exposed back, a tour group shows up in the Media Lab unannounced. A crowd of kids, maybe fifth or sixth graders, approaches the robot. Nexi is tracking their faces when one of the boys gets a little too close. The robot's eyebrows swivel inward. The eyelids narrow as the head tilts down. And the worm motors that control Nexi's fingers whine like electric drills as its fists clench.

"Whoa!" the kid in the lead says, and they all backpedal.

"Is it getting mad?" one girl asks the researchers.

Then Nexi's face softens and, instantly, they're laughing.

"So do you give robots emotions?" another girl asks.

I remember something Breazeal told me earlier: that for kids who grow up around robots, the uncanny valley could be irrelevant and *The Terminator* little more than a quaint story. Vulnerable or not, children interact with these machines differently. Understanding the limits and strange potential of robotics might be as simple as letting them meet the models most like them—the ones built to live at their sides. Maybe Nexi could act as that first, limited exposure, a vaccine against the wild fears and warped perceptions the rest of us have grown up with.

The kids provoke Nexi's anger response again, laughing more this time. When its eyebrows level, the lead boy jabs his friend and points at the robot's impassive face.

"It's smiling at you! It's smiling!"

EDITOR'S NOTE

Dr. Breazeal is still hard at work at the MIT Personal Robots Group. She has used free-form human feedback to train Nexi to perform multiple behaviors—no computer programming necessary.

Kobian's latest adventure has been learning how to make humans laugh. Researchers presented a paper at the 2014 IEEE International Conference on Robotics and Automation in which Kobian experimented with such techniques as "blue jokes," "imitation," and "self-deprecating humor."

As of 2014, Zeno has undergone a series of upgrades and now stands 27 inches tall, has an increased range of motion with 37 degrees of freedom, and can converse in 26 different languages. —DHW

THE MACHINES ARE WATCHING

→ Las Vegas casinos are incubators of the world's most advanced surveillance tech. And the spy gear that helps Sin City has taught everyone from government to big banks how to snoop more effectively.

BY **MICHAEL KAPLAN**

This article was published in *Popular Mechanics* in 2010.

T IS 2 AM INSIDE THE BUNKER-LIKE surveillance room at the Mirage Resort in Las Vegas, but 28 wall monitors show there's still plenty of action down on the floor. A surveillance worker we'll call Tom logs in and starts the graveyard shift, taking an overhead tour of the 100,000-square-foot casino. Using a joystick, keypad and three desktop screens, he surveys video from some of the 1,000 ceiling cameras.

Tom is a table-games specialist, so he starts by scrutinizing a few poker hands, then sweeps over medium-stakes blackjack and watches a busy craps table. Nothing looks unusual until he stops at a baccarat game in the high-limit room, where betting minimums start at $100 per hand. He focuses on a young Asian man in a white suit who keeps his hands curiously positioned. Sometimes they cover the cards in front of him; at other times they rest on the side of the table. Suddenly, the man sweeps one hand up along a lapel of his jacket.

Like many gamblers in Las Vegas, the man presented a players card, the equivalent of a customer-loyalty card, to the dealer before buying into the game. Through these cards, the casino monitors the play of guests and dispenses complimentary goodies accordingly (risk enough money, and you may wind up in a villa with a butler). The card enables Tom to retrieve a profile of the player: his name, date of birth, address, amounts won and lost on previous visits and other data.

Tom checks the player's long-term success rate at baccarat: He's a stone-cold loser. Common sense suggests that his poor record should exonerate him. Playing a hunch, Tom uses an internal search engine to correlate every player and dealer that the suspect has gambled with at the Mirage. One name repeats—a big winner, also Asian. On this trip alone, he's ahead hundreds of thousands of dollars, and he happens to be playing right now, at the same table as Mr. White Suit.

Less than an hour later, Tom makes the call. He is convinced that the fellow in the white suit is not rubbing his lapel but dipping his finger inside his jacket. He is swapping cards in and out of the game, a tactic known as hand mucking. Capitalizing on baccarat's simple rules, which allow gamblers to take the side of player or banker, Mr. White Suit loses minimal wagers while his confederate wins large ones from the casino.

THREE *TYPES OF CAMERAS FEED THE VIDEO WALL IN THE MIRAGE'S SURVEILLANCE ROOM. FIXED-FIELD-OF-VIEW UNITS FOCUS ON TABLES, MOTORIZED PAN-TILT-ZOOM CAMERAS SURVEY THE FLOOR, AND 360-DEGREE CAMS TAKE IN AN ENTIRE AREA.*

When the winning conspirator attempts to cash out his chips, guards detain him. Other guards hustle the mucker from the table. The cheater tries to break free, then drops to his knees and eats the card that he had slipped inside his jacket. He may have swallowed the evidence, but the casino's digital ceiling cameras have captured all of his illicit actions.

Soon after this incident, the Mirage outfitted its baccarat tables with a system known as Angel Eye. A scanner hidden in the shoe—the plastic case out of which cards are dealt for multideck games—reads invisible bar-code strips on the cards. "Angel Eye identifies the cards as they come out and conveys that information to the dealer," director of surveillance Ted Whiting says. If a player swaps in a card, the dealer knows. "That one change put card muckers out of business here."

NTER A MAJOR LAS VEGAS CASINO, and you might as well be walking into a complex computer built to study your relationship with money, your motivation for gambling, even your taste in food. Cameras capture your every move, software calibrates your play, and regressive-analytic applications (like those used on Wall Street to predict a stock's future) estimate your long-term worth to the casino.

Given the wild bets taken recently by investment banks, the overlap of gambling and financial technology may not be surprising. But the innovations pioneered for Las Vegas surveillance rooms have significance and applications that reach a lot further than a trading floor. According to Dave Shepherd, former executive director of security at the Venetian Resort Hotel Casino, Las Vegas, is an ideal proving ground for innovations that eventually end up in airports, shopping malls, and government agencies. "There is no Underwriters Laboratory for security technology," says Shepherd, who serves on a casino-focused council affiliated with Homeland Security. "Casinos use the earliest versions of security and surveillance devices. People in other industries see how they work, and those people come up with fresh applications for the technology."

Vegas's gaming industry, after all, has the resources and incentives to be a pioneer in surveillance tech and data mining. "Casinos employ the most talented cryptographers, computer security experts and game theorists," says John Pironti, chief information risk strategist for Archer Technologies, a Kansas based company that specializes in data protection. "Casinos are vulnerable and have a vested interest in being innovative."

A modern Vegas property is a microcosm of a wider world, with restaurants, a hotel, entertainment venues, retail shops and a sophisticated system of currency exchange. It's all in a highly controlled environment where customers eagerly volunteer personal data for a chance at comps. As a result, casinos maintain a treasure trove of information on customer behavior that most marketers would die for. Players cards and gambling, in general, are opt-in propositions. The casino industry is highly regulated, and the watchful tech is not only legal but, in many cases, mandated. Still, the opaqueness of the programs is a cause of concern for some privacy advocates. "Why should casinos have secret files on their best customers?" asks Marc Rotenberg, executive director of the Electronic Privacy Information Center (EPIC). "People should know the information that casinos gather on them."

Digital data has a long memory, and effective surveillance technology spreads fast. The software that measures your gambling skill at the blackjack table today could be gathering data for your performance review at work tomorrow.

Paying close attention to customers is as much a security concern as it is a marketing opportunity for casinos. From the moment you place your first bet with your players card, the casino starts paying attention. "That financial transaction feeds into a data-warehousing platform," says David Norton, chief marketing officer of Harrah's Entertainment. The most direct interface with the system is a modern slot machine. These days most slots are run by computers, and until recently, all of these computers have been self-contained machines. To make adjustments on standard slots, attendants have to stop play, open the housing, and swap out chips, a time-consuming process that reduces profits for the casino. The Mirage's soon-to-open sister property, Aria Resort &

MONITOR # 16
MONITOR # 17
1001
28SEP2009 12:02.54
Ikegami
MONITOR # 13
1289
28SEP2009 12:02.54
CEC-CMH19A
PROFESSIONAL DISPLAY SYSTEM
CCTV COLOR MONITOR
MONITOR # 11
0343 CREDIT
28SEP2009 12:02.54

MONITOR # 18
MONITOR # 19
Ikegami
MONITOR # 14
1304
28SEP2009 12:02.54
CEC-CMH19A
PROFESSIONAL DISPLAY SYSTEM
CCTV COLOR MONITOR
MONITOR # 12
Dealer must draw to 16
0063
28SEP2009 12:02.54

Casino, however, will be the first casino in Las Vegas outfitted with server-based slot machines. That means Aria's one-armed bandits will run off a single computer, allowing supervisors to alter machines simply by pushing backroom buttons that can change games, odds, and limits to suit the player or the situation. If a player is in town for the National Finals Rodeo, the slot machine could load up a game with a rodeo theme and alert the player when certain comps kick in or provide the showtimes of events he might be interested in. It'll even wish him happy birthday.

All the personal attention may seem flattering, so long as the casino values your business. But what about those people who are viewed as undesirable? At the Venetian Resort Hotel Casino, special software allows security workers to enter a suspected bad guy's characteristics (a mustache, say, along with a forearm tattoo and a habit of lurking around roulette tables). If there is a

→ Data Jackpot

License-Plate Reader

Many casinos know who you are before you even walk through the door. At the self- and valet-parking areas of the Mirage, for example, cameras scan the license plates of vehicles as they enter. Pictures of every plate are then run through optical character-recognition software. If your plate matches a database of undesirables, the security personnel may hand back your keys and suggest you take your business elsewhere.

Beyond Vegas

License-plate scanners are now commonly deployed in police patrol cars to check traffic for suspect vehicles.

Eye in the Sky

Thousands of cameras built into the ceiling can cover more than 80 percent of a casino. Computer-vision systems automatically scan for suspicious activity on the floor (people congregating in odd areas, unattended bags) as well as at the tables (dealer errors, cheating players).

Beyond Vegas

Similar systems are also used by airports to watch for potentially dangerous activity, as well as by retailers such as Best Buy, which uses the technology to monitor traffic patterns in stores and to harvest data on customer shopping behavior.

Smart Tables

Several systems kick in once you get to the table: Cards printed with invisible bar codes discourage deceitful players from swapping in fakes; non-obvious relationship awareness (NORA) software determines if you share enough background data with the dealer to be suspected of collusion; and analytic programs determine your skill as a player.

Beyond Vegas

Developed for casinos, NORA technology is now used by Homeland Security to look for ties between suspected terrorists. Banks and insurance companies also use NORA to sniff out relationships between customers.

visual match from the casino's database, it pops up on the screen, along with identification data. "We have a lot of coverage, a lot of cameras, a lot of information," says Dan Eitnier, head of surveillance at the Venetian. "A couple of years ago we had a collusion situation, and by finding the suspected dealer's car-loan application in our file on him"—the lender had asked the Venetian to confirm his employment there—"I saw that he gave one of his frequent players as a reference." The scheme unraveled from there, and both men were busted. New algorithms have elevated this type of on-the-spot background check to a Vegas art form. Non-obvious relationship awareness (NORA) software allows casinos to determine quickly if a potentially colluding player and dealer have ever shared a phone number or a room at the casino hotel, or lived at the same address. "We created the software for the gaming industry," says Jeff Jonas, founder of Systems Research &

LAS VEGAS'S GAMING INDUSTRY INVESTS IN THE BEST SURVEILLANCE AND BEHAVIORAL MONITORING TECHNOLOGY IN THE WORLD. BUT CASINOS AREN'T THE ONLY ONES INTERESTED IN HIGH-TECH SNOOPING. HERE'S WHO'S LOOKING AT YOU.

RFID Chips

Money talks in Vegas, but your chips speak in code. Some casinos, such as Wynn Las Vegas, have high-frequency radio transceivers hidden in the chips. The technology can be used to confirm that the chips are legit and can also be used for real-time accounting, so that management knows where the money is at all times.

Beyond Vegas

Tiny, cheap RFIDs are so pervasive that millions of people carry at least one of these trackable transceivers on themselves at all times, in the form of a corporate ID, contactless credit card, or toll-collection pass.

Networked Slots

Server-based computerized slot machines allow casino management to change games and set odds remotely, then push games out to each machine from a centralized location. When used with a loyalty card, the game can track betting patterns and deliver a customized game to the player.

Beyond Vegas

Slot machines all operate using a complex algorithm known as a random-number generator. And the same type of program that determines jackpots is also useful for high-tech cryptography, which protects government secrets via encryption.

Cashier's Window

The cashier's window is the last line of defense against those who try to take advantage of the house. Automated document scanners can determine if an ID is valid before a cashier dispenses a credit card advance for chips. If things don't match up, an automatic call goes to security before the perp or cashier even realizes there's an issue.

Beyond Vegas

Real-time document verification scanners are used to instantly check the authenticity of IDs at border crossings, banks, and nightclub doors.

Development, which originally designed NORA. The technology proved so effective that Homeland Security adapted it to sniff out connections between suspected terrorists. "Now it's used as business intelligence for banks, insurance companies, and retailers," Jonas says.

According to EPIC's Rotenberg, any industry that collects so much data on its customers is at risk for a computer security breach. "Even if casinos have no interest in using their information for any purpose other than the intended one, things don't always go as planned." Especially, he points out, since security teams at competing casinos often share information.

With all the data collection and camera monitoring going on in casinos, a sense of gambler's paranoia is understandable. But it's worth remembering that the same technology that protects the house could end up protecting you. Casinos are tempting places for pickpockets; customers stroll the floors with cocktails in their hands and thousands of dollars in their pockets. Some of the sexiest-sounding software—facial-recognition systems that promise to set off alarms as soon as a known criminal enters the property—is still too primitive to be useful. However, more reliable analytic software is employed in casinos such as the Mirage to monitor video feeds for suspicious activity—someone hiding in a stairwell, for example, or a purse left unattended too long.

HE MOST ADVANCED SURVEILLANCE TOOL in the gaming industry is focused on the blackjack tables at Barona Resort & Casino in Southern California, where management aggressively tests new technology. The system, called TableEye21, was created by Canadian computer engineer Prem Gururajan to profile and rate players according to skill.

TableEye21 uses overhead video cameras and video analysis software, and can track information from casino chips embedded with radio frequency (RFID) transmitters. The system quickly identifies "advantage" players who can cost casinos profits. These gamblers use legal strategies such as card counting and shuffle tracking, in which the player watches for clumps of favorable cards. Gururajan says TableEye21 will be coming online soon at a Vegas casino, and surveillance specialists are enthusiastic about the product. "You get a printout of the player's skill level, how much you can expect to win from him, and whether the dealer is making errors," Gururajan says. "Since the system tracks the player's bets, the casino knows exactly how good a customer the player is."

Sometimes, casino monitoring can go too far. A few years ago, a product called MindPlay hit the market. Fourteen tiny cameras photographed cards as they came out of the blackjack shoe. The system's software executed a quick bit of analysis and notified dealers, in real time, whether shoes were cold or hot—that is, when the remaining cards favored players. "That would be a good time for the casino to come up with an excuse to shuffle," says veteran security director Arnie Rothstein. "Players found out about it and complained to the Gaming Control Board." The product, according to its manufacturer, is no longer in use.

Inside his plushly carpeted surveillance lair at the rococo Venetian, Dan Eitnier inspects the flat-screen monitors on the walls. He acknowledges that technology runs both ways in the gaming business: The operators aren't the only ones who capitalize on cheaper bytes and easy access to data.

Eitnier admits that all casino games are vulnerable. Enhancements in technology have simply added another layer to the endless cat-and-mouse game played by those who are paid to protect casinos and the renegades who get rich by out-thinking the protectors. "Whenever new technology is introduced, you always have people out there who want to beat it," he says. Cheaters buy and dissect slot machines, angle-shooters analyze automatic shufflers in search of patterns, and card counters continue to stymie facial recognition. "They find weaknesses in the technology, and then we come up with new technology that they have not yet figured out."

Eitnier leans back in his chair and keeps his eyes on the monitors. He smiles. "Of course," he says, "without those people trying to beat everything, I wouldn't have a job."

EDITOR'S NOTE

These types of surveillance enhancements have proliferated into day-to-day society. If you wonder whether the machines are still watching, just walk into your living room and glance at your XBox One— it will recognize your face via the Kinect interface and log you in automatically. —DHW

THE ROBOTS ARE COMING . .

One suburban homeowner's quest to automate his lawn into submission.

GLENN DERENE

This article was published in *Popular Mechanics* in 2013.

OR SOME PEOPLE, MOWING THE LAWN IS a meditative experience—a chance to tune out while getting a little exercise walking behind the lawn mower, inhaling the scent of freshly cut grass. It's good old-fashioned domestic man's work, like your father did before you, his father before him, and so on. Well, not me. I hate mowing the lawn. It's a numbingly repetitive, sweaty, noisy waste of time. My father hated it, too. And I'm pretty sure his dad did before him.

Tell you what I do like, though: robots—love 'em! In fact, I would gleefully surrender every thankless bit of home landscaping to an automaton. So I decided to see if I could piece together a system wherein my lawn essentially would take care of itself. Yes, I could have hired a landscaping crew, but to me that was a dodge. I didn't want to pass off my dirty work to someone else. That's the beauty of robots—one day they may take over the world, but for now, they get the grunt work.

And, it's worth mentioning, I wanted a beautiful lawn—green, lush, carpet-like—something my family and I could really roll around on during a midsummer day. I just didn't want to sweat for it. The good news is that, for mowing, there's already a robot solution—a couple of them, in fact. Honda sells the Miimo; a company called LawnBott offers a variety of, well, lawn bots; and Friendly Robotics has a bunch of really friendly looking mowing robots. All of these systems seem pretty similar and promise essentially the same thing: to tame your turf with a minimum of human oversight.

TO MOW MY LAWN

I called up Husqvarna, a company with a long history in the grass-cutting biz. Husqvarna also has deep experience with robotic lawn mowers; it introduced the first consumer model in 1995. Now it sells two: the Automower 230 ACX and the Automower265 ACX. A few weeks after my call, I got a big box with a 265 ACX and an appointment with company representatives Quinn Derby and Gent Simmons. They arrived a few days later, surveyed my postage-stamp-size lawn, looked at the box from their company, and concluded that the 265 was complete overkill. But the machine was there, so they decided to install it anyway.

Now, two animal analogies are useful for understanding the operation of the Automower. First, the machine works like a sheep, roaming about your lawn at random, nibbling away at the blades of grass in small increments. Second, the Automower is prevented from leaving your lawn in much the same way that a dog can be contained within an electric fence. That's what Derby and Simmons installed on my lawn—a fence for the robot. They laid down a low-voltage wire that creates a mild electromagnetic field so the robot can sense when it has reached the prescribed boundary, then back up, turn randomly, and proceed in another direction.

Derby informed me that my lawn was an unusual one for the Automower. That's because I really have two lawns: a small one in the front of the house, and a larger one in the back, separated by a fence and a patio. I could tell that Derby is by nature a problem solver, as he described several ways he could attempt to string the wire to allow the robot to travel from my back lawn to the front lawn. But I imagined the poor, confused machine attempting to nudge aside patio furniture along the way, and I told Derby to fence them off separately—occasionally I'd just pick the thing up and move it to the front lawn myself.

The two men installed the perimeter wire and base station and then gave me some brief advice on programming the mower—1 hour, for four days a week, would keep the back lawn neat and tidy; 1½ hours once a week would handle the front.

It was bliss. Every Monday, Tuesday, Wednesday, and Saturday at 1 o'clock, the mower would back out of its charging dock and quietly go about its business until 2 o'clock, when it would find its way back. The mower is so quiet that I considered rescheduling it for the middle of the night. My 4-year-old

son, Owen, protested. Watching the mower cruise across the lawn had become one of his favorite activities. On Sundays, after setting it on the front lawn, I would uncap a beer and hand Owen a juice box, and we would sit in the rocking chairs on our porch and watch the Automower go—ironically, paying more attention than ever to our lawn.

It was easy to anthropomorphize the thing. We found ourselves talking to it ("You missed a spot! Look out for that tree!"), dogs barked at it, and the sidewalk passersby found it fascinating. "It's so cute!" said the woman from two houses down. "It's like a Roomba for your lawn." Then there were those who found it ominous: "Aren't you afraid it's going to cut your toes off?" said one kid, who nearly fell off his bike when he first saw it. The answer was no. The Automower is loaded with sensors designed to prevent injuries. In addition to collision detection that makes it back off the instant it touches anything, it has a lift sensor that stops and retracts the recessed blades if you try to pick it up. According to Derby, the biggest problem Husqvarna has seen with the Automower is kids trying to ride on top of it. It's enough of an issue that the mower now ships with a sticker warning against it.

Robotic mowers are a rarity on American lawns, though they are popular in Europe, where landscaping services are particularly expensive. But automated seed and fertilizer spreaders don't exist anywhere. I know, because I checked. So I decided to create such a machine myself.

Well, not entirely by myself. I called up Randy Sarafan, technology editor at Instructables and an occasional *Popular Mechanics* collaborator. He had previously roboticized an RC monster truck, which I figured would make a good platform for a spreader. I thought we could mount a handheld spreader to the top of Sarafan's robot truck, then connect a belt drive from the axle of the truck to the crank of the spreader. Every time the truck moved, it would spread the good gospel of grass. I bought an RC truck to test it out. The truck struggled to move itself through thick grass, so there was no way it was going to power a spreader, as well. Sarafan suggested starting from scratch with an Arduino-controlled, dual-motor, tracked platform with an independent third motor driving the spreader. He could pull that off easily, he said.

Guidance would be the tough part. At first, I figured the little spreader could use the same guide wires as the Automower, but a quick talk with *Popular Mechanics* senior home editor Roy Berendsohn exposed that as a bad idea. It's okay for a mower to move randomly within a defined space, he explained—you can't overmow a lawn as long as the blades aren't set too low—but letting a seed-and-fertilizer spreader loose like that would be a waste. Sarafan designed the robot with a sonar proximity detector and a clever set of instructions. My main patch of lawn is rectangular and bordered on two sides by a fence, so the robot would start at one edge and count the number of revolutions of its wheels until it sensed the fence in front of it. Then it would turn around and progress back the same number of wheel revolutions until it reached the other side of the lawn, and repeat. He tested the thing at the Instructables headquarters in San Francisco, then shipped it to me on the East Coast.

It was quite a ceremony at the Derene house when I first brought out the seed bot. My wife and Owen and his baby brother, Elliot, all came out to see its maiden voyage. I had high hopes for the little machine—the only one of its kind on earth. I filled the hopper and flipped the robot on, and off it went, rumbling slowly on its treads with a spray of seeds shooting out in front of it. But our excitement was short-lived. The little bot made it halfway to the fence, stopped, turned, and progressed back at a cockeyed angle, lumbering diagonally across the lawn toward my wife's Volkswagen in the driveway. "Daddy, stop it!" cried Owen. "It's going to hit mommy's car!"

I ran after the wayward bot, which was still flinging grass seeds all over the place, picked it up, and turned it off. In its undisciplined way, it had spread seed on about one-quarter of my lawn. Not a complete failure, but not quite a success, either. Sarafan and I would have to make some code corrections and maybe adjustments to the sonar. But that would come later. Right then, I did what any good husband, father, and homeowner would do. I got out my old conventional seed spreader and finished the job myself. Total robot lawn domination would have to wait.

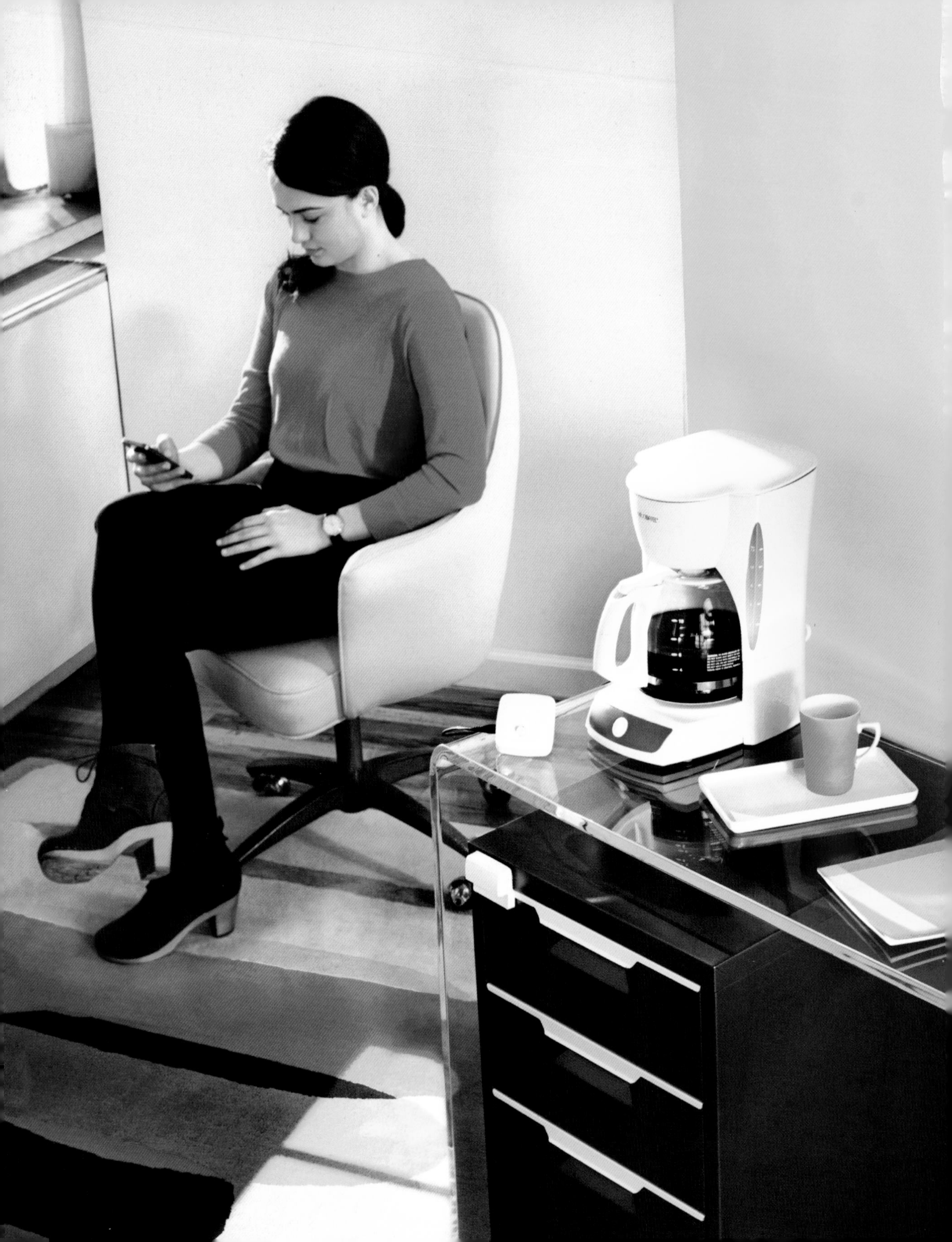

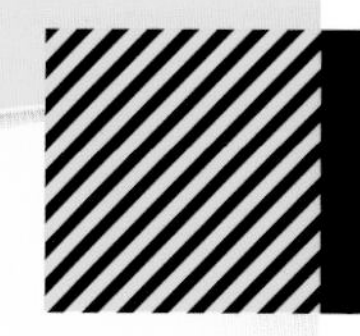

I, COFFEEPOT

How the Internet of Things revved up my apartment—and sort of creeped me out.

BY **ALEXIS SOBEL FITTS**

This article was published in *Popular Mechanics* in 2014.

FEW WEEKS AFTER I GAVE MY APARTMENT a home-automation makeover, I found myself in a strictly 21st-century pickle: I was naked and shivering, stranded post-shower in my bathroom, trying to plot a route past a motion-activated camera to the freshly laundered towels I'd left by the bed. The camera was only meant to shoot pictures of trespassers, but the trigger that shuts off the system when I'm home is my smartphone–and I'd allowed the battery to run down. From technology's point of view, I was an intruder in my own apartment.

In the end I gave up. The camera captured my awkward, dripping sprint, and I deleted the photos later. To avoid a repeat of the problem, I set up a key-chain fob–one with a nearly infinite battery life–to let the system know when I am home. It was one more step along my home-automation learning curve; as I was finding out, merging your digital and physical worlds requires some troubleshooting.

SMARTTHINGS *IS ONE OF A NEW CROP OF HOME AUTOMATION SYSTEMS THAT CAN CONTROL DEVICES REMOTELY FROM A SMARTPHONE. COMMON GIZMOS MIGHT INCLUDE: A COFFEEPOT AND ITS SMART OUTLET TRIGGER; SONOS MUSIC SPEAKERS; A SENSOR THAT ALERTS YOU WHEN A DRAWER IS OPEN; AND A WATER-LEAK SENSOR.*

HE INTERNET HAS LONG SINCE TRANSFORMED the way we communicate and share information. We deposit checks from our phones, send messages through Snapchat, and carry around hundreds of books on our Kindles. But now the digital is creeping into our lives in a new way, letting us remotely monitor and control physical objects, from thermostats to door locks to electrical outlets. The Internet of Things, which could eventually incorporate billions of objects, is starting to go mainstream.

Home-automation systems existed in the 1970s, but clunky remotes and steep prices limited their appeal—plus, the Internet didn't exist back then. More recently the rise of the smartphone has transformed the technology. It provided an ever-present and digitally connected new remote, allowing smart homes to get past the clap-on lights of bachelor pads of yore. Today's systems vary in complexity: You might prefer a simple setup (a single thermostat) or a more intricate one (a blanket system that controls all your lights, the sprinklers on your lawn, and the music in your living room).

I wanted to see if using technology to control my home would make my life easier or complicate things instead. Several players are competing in this market; I chose SmartThings, an ecosystem known for being highly customizable. The starter kit comes with various "things"—a motion detector, sensors that attach to drawers, and two key-chain sensors. There's also a hub that grabs the information from all those components and beams it through the cloud to a mobile app.

First, I turned my phone into a presence tag, which is a GPS automated ID that lets the SmartThings app establish a virtual perimeter around my place. When my cellphone moved into that geofenced zone, my apartment knew I was home. SmartThings is open source, and a number of third-party companies offer products that can sync with the system. I connected a set of Sonos speakers to play music at predetermined times. I installed a camera to text me photos of trespassers, and then attached a sensor to the drawer where I keep my documents. I plugged my lamps and a coffeepot into smart outlets so they could turn on and off automatically. My Hue LED lightbulbs went online too: They now change colors with the flick of a finger on my smartphone.

FTER ALL THE SENSORS WERE UP AND RUNNING, I pressed Good Morning on the SmartThings app. "Good morning," my house replied. (So friendly!) "I changed your mode from Night to Home as you requested." All my lights went on, and my coffee began brewing. I sipped a cup while listening to the Beyoncé single I had asked SmartThings to play when I got out of bed. It felt like having an attentive, yet invisible, butler.

SmartThings wasn't built for amusement, though. The company's founder, Alex Hawkinson, conceived of the platform after a broken pipe flooded his family's vacation house leading to a $100,000 repair bill. "My daughter was streaming Netflix," he recalls. "There was all this bandwidth in the air, but none of it was directed at the problem of what was going on in my home." Hawkinson has now installed about 200 smart devices in his home. And, yes, they include moisture sensors that can send text messages when water is detected.

The SmartThings blog is filled with stories of homes saved from fires, hurricanes, and leaky pipes. But the typical user is looking primarily for safety and security (surveillance cameras) and energy savings (turning off lights and appliances), along with a bit of convenience (morning coffee).

The settings can take some time to fine-tune. The day after I set up my system I had a friend over for dinner, but needed to dash out to buy butter. When in line, I received a text from her: "All of your lights just went out!!!" I hadn't thought to program a scenario for guests when my phone and I aren't present. Hawkinson says he programmed his family's kitchen lights to turn on whenever the system detected motion, but then his wife walked down in her underwear and almost flashed the neighbors. He reprogrammed the lights to work only after 7 am.

Embarrassing snafus simply take some experimentation to overcome. Privacy issues, on the other hand, present a more enduring concern. Home-automation systems turn your physical behaviors into digital information—that's how they work. They can monitor when you get home from work, whether you're alone, and what time you go to bed.

Lee Tien, a senior staff attorney at the Electronic Frontier Foundation who specializes in privacy laws, worries that data-protection problems will multiply as the Internet of Things matures. "The home has traditionally been the most sensitive place that we recognize in both our legal system and in the social sense," he says. "All of these systems put a very large aperture inside that wall of privacy."

Last summer a hacker used a blackout command to successfully break in to a set of Hue bulbs. The consequences were hardly catastrophic–the lights shut off. But the possibilities are disturbing. One night I forgot to set my app to Good Night mode, which would have cut my security camera, before I went to sleep. I awoke to 17 texts: "There is motion in the Bedroom and photos have been taken at Alexis's Apartment." I scrolled through shots of myself in bed, sleeping.

SmartThings has strong privacy protections, encrypting data as it's captured by the hub and again as it's passed to the cloud. No one else had access to those photos, but the experience still felt creepy. If a hacker were to break in to my SmartThings app–or if someone got hold of my phone–he could track my movements in my home. And, Tien points out, even encrypted data can be obtained by a court order.

Right now, though, those concerns seem largely theoretical–and I've decided to keep the SmartThings system running. For one thing, my automated lights have cut my electricity bill by 10 percent. Of equal importance: The coffee is ready when I wake up, and so is Beyoncé.

→ Omniscient Objects

I've started to welcome more and more smart gadgets into my home, like thermostats, smoke detectors, locks, and lightbulbs. But I'm concerned: These devices collect data on my habits and power usage. Who has access to it–or can demand access to it?

An increasing number of objects– in the home, the car, and even the office–are being embedded with sensors and acquiring the ability to communicate. They help make up the so-called Internet of Things, a rapidly growing product category, with items that range from gratuitous (Internet-connected toothbrushes) to lifesaving (water-leak sensors that can help you catch floods in your home when they start). Google, for one, apparently expects it to be an enduring trend: In January the company paid $3.2 billion to buy Nest Labs, which makes smart thermostats and smart smoke alarms. Competing product lines include Lowe's Iris system and a range of modular sensors made by SmartThings.

The way the companies deal with user data varies with the architecture of their products and how they link up to the Internet. For example, Nest devices collect and process data locally, but the data is sent to the servers periodically to be analyzed for feature improvements and energy reports. The company says only a handful of members of an internal-quality team see any information, and it's a mere sampling of what's generated by all the Nest products in people's homes. Of course, users themselves have access to a snapshot of their data via the company's mobile apps.

In contrast, most SmartThings products rely on cloud services to operate. For instance, a user can receive a push or text notification if a water leak is detected, or simply when a child gets home from school. The company's CEO, Alex Hawkinson, says that SmartThings analyzes some of the data to help it improve its products and services, but that the information is anonymized.

"There's no way for us to know, for example, that somebody in New Jersey has an 8-year-old daughter named Katie, just because they use our system," Hawkinson says. He promises the company would try to resist any government request to access the data, though it hasn't come up. "We believe that consumers should own their physical graph, and all of the data that results from that physical graph," Hawkinson says.

Nest cofounder and engineering vice president Matt Rogers says his company has never received a government request for customer information. "Our contract with our users is that we will keep their data private," he says. The company has publicly stated that any data sharing that happens with Google, its new owner, will be transparent and require the user's permission. But Nest has also hinted that deep-data integration with Google could be coming soon. In the future, conceivably, information gathered from one's quantified home could be used for commercial purposes.

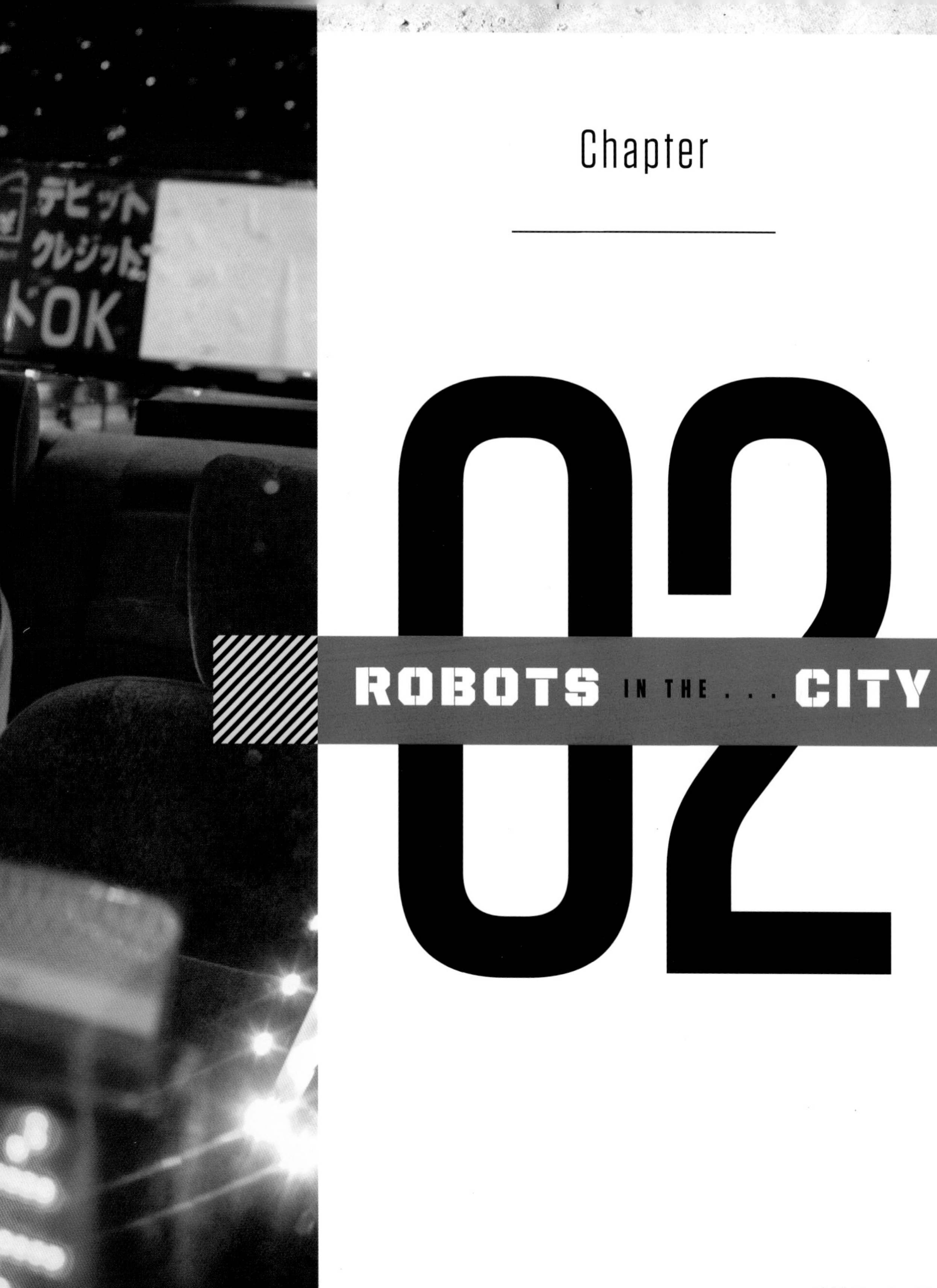

Chapter 02

ROBOTS IN THE . . . CITY

ESCAPE FROM THE LAB

→ The robot looks relaxed behind the wheel of the all-terrain vehicle—one three-fingered hand gripping the wheel, the other clamped to the roll bar overhead.

BY **JOE PAPPALARDO**

This article was published in *Popular Mechanics* in 2014.

NE MECHANICAL FOOT HOVERS OVER THE accelerator. The bipedal robot has the aspect of a laid-back teenage driver, flouting the rules of every driver's-ed teacher in the world. The casual pose is meant to keep the 209-pound machine from sliding in the seat as five young Japanese engineers from Tokyo-based Schaft, wearing blue vests and white hard hats, check power connections and run diagnostics on a laptop. It's almost time for S-One–which looks like a minifridge with unnaturally long arms and strong legs that bend backward at the knee–to go for a ride.

There are hundreds of spectators at Homestead-Miami Speedway in Florida this December weekend. But instead of NASCAR or Indy Car races, the crowd, pressed to a chainlink fence, is waiting to watch the teleoperated bot drive a Polaris Ranger XP 900 down a winding course. The 250-foot route is hemmed in with lane dividers, and empty plastic barrels are stacked two high at each of the six turns. Officials from the Defense Advanced Research Projects Agency (DARPA), including director Arati Prabhakar, join the media on the other side of the fence to get a closer look before S-One powers up. Prabhakar is a diminutive woman in a red baseball cap, her trademark white hair poking out the bottom and a matching red vest that says DARPA Director on the back. She eyes S-One with interest and some suspicion. "People here are pretty comfortable with a robot behind the wheel," she says of the swarm of technicians and media surrounding the all-terrain vehicle.

The machines have center stage, but the drama is human. The robot is only a tool, controlled by human masters. Housed in a garage several hundred yards away, the operators have only S-One's cameras and laser sensors with which to perceive the outside world.

Finally, the humans on the route clear a space and the robot is ready to roll. Its right foot pushes down on the accelerator, and the Ranger lurches forward. At the first turn, the foot lifts and the ATV halts–there's no need for

NAME: ***ROBOSIMIAN***

TEAM: ***NASA'S JET PROPULSION LABORATORY (JPL)***

HEIGHT: ***HEIGHT: 5 FEET 5 INCHES***

WEIGHT: ***238 POUNDS***

THIS ***ROBOT WALKS LIKE A SPIDER BUT CAN CLIMB AND MANIPULATE TOOLS LIKE AN APE.***

ROBOSIMIAN

S-One to tap the brake, since the vehicle uses engine braking to automatically slow down. This isn't NASCAR: Six stop-and-go turns take 20 minutes. Applause and cheers fill the morning air when the vehicle crosses the finish line.

The Schaft team has become the first to complete the drive, one of eight tasks at the DARPA Robotics Challenge (DRC) trials, held late last year. Sixteen teams from universities, companies, and NASA have gathered at the speedway to compete in the most demanding robotic competition ever staged. Each event is designed to prove that robots can help people in the aftermath of disasters. DARPA intends to award $1 million for each of the top eight finishers, who will appear in the finals.

The DRC winner will take home $2 million, but those competing believe the stakes are larger. If the robots here perform well, they could jump-start a lucrative industry and reimagine the relationship between man and machine. "Mobile robotics is where the dot-com boom was during the 1990s," Eric Meyhofer, lead technician of Team Tartan Rescue from Carnegie Mellon University, says. "We're starting to see real general interest in this market. We're lucky to be smack-dab in the middle of it."

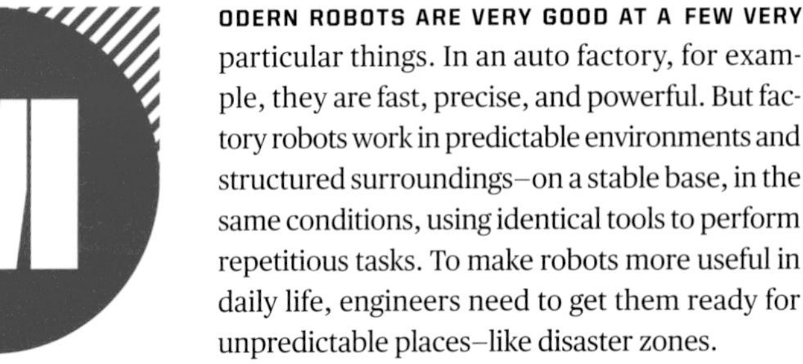

MODERN ROBOTS ARE VERY GOOD AT A FEW VERY particular things. In an auto factory, for example, they are fast, precise, and powerful. But factory robots work in predictable environments and structured surroundings—on a stable base, in the same conditions, using identical tools to perform repetitious tasks. To make robots more useful in daily life, engineers need to get them ready for unpredictable places—like disaster zones.

DARPA modeled events at the DRC on conditions at the Fukushima Daiichi nuclear plant in Japan during its 2011 meltdown. Throughout that disaster, wheeled robots entered radioactively contaminated areas but could only transmit video images. If a robot had been able to open valves to vent hydrogen gas from the reactor, it might have prevented subsequent explosions. A robot capable of commandeering fire trucks abandoned in Fukushima's contaminated zone could have refilled water in the ultrahot spent-fuel pools. "The world is built for human beings," Gill Pratt, the DARPA program manager who spearheaded the robotics challenge, says. "Robots need to operate in that environment."

To take robots out of the labs and factories, Pratt based the challenge around basic skills. Each team's robot must walk across uneven terrain, climb a ladder, pick up a human tool and use it to cut through drywall, connect a fire hose, and turn valves. These are all tasks that a human being can easily do, and faster than a robot, but not in a radioactive or chemically toxic environment. "You won't see robots racing to the rescue," Pratt says. "You'll see robots being deliberate to the rescue."

DARPA added another wrinkle by limiting communications between the robot and its operators. Every other minute, a black box in each team's garage disrupts the signal to the robot, cutting the bandwidth to a narrow sliver. Such disrupted communications were a hallmark of Fukushima, and will likely be the case at future disaster sites.

The crucial engineering challenge at the DRC is to field a single robot that can perform all eight tasks. A door-opening robot is easy to envision and program, but making one that can also drive an ATV and climb a ladder adds a lot of complexity to the challenge. Most of the teams here are using man-size and -shaped robots that can navigate environments built for humans—think of the height of stairs, the size of doors, and the location of cabinets. This also benefits the operators, who can more easily imagine what a robot is doing, if it is built like a human. But a few DRC robots have alien aspects: NASA's Jet Propulsion Lab, creator of the Mars rovers, operates a 238-pound, four limbed creation called RoboSimian that uses seven actuators (joints) in each limb to brace itself when a limb applies force. Each of the six legs of the spindly, 150-pound Chiron, made by Utah-based Kairos Autonomi, can also perform manipulation tasks.

The most common robot at the DRC is the Atlas, a 6-foot, 2-inch, 330-pound humanoid built by Boston Dynamics. DARPA purchased six for teams that earned their way to the DRC by writing code and defeating rivals in a virtual contest. Atlas has cameras and a scanning laser radar (lidar) where a human's face would be. Given the DRC's restricted bandwidth, the cameras provide only grainy views of the surroundings. The flickering lidar forms a rainbow of points on the operators' screen, giving them the long distance perception lacking in the cameras. The

TOP: *AN ATLAS ROBOT NAVIGATES UNEVEN TERRAIN.*

MIDDLE: *S-ONE, FROM THE JAPAN-BASED FIRM SCHAFT, CLIMBS A LADDER USING LEGS THAT BEND BACKWARD AT THE KNEE.*

BOTTOM: *THE DARPA ROBOTICS CHALLENGE AIMS TO PROVE ROBOTS CAN ASSIST IN THE AFTERMATH OF DISASTERS. MEMBERS OF TEAM MIT, INSIDE A GARAGE AT THE HOMESTEAD-MIAMI SPEEDWAY IN FLORIDA, PREPARE TO TELEOPERATE AN ATLAS ROBOT DURING ONE OF EIGHT EVENTS.*

operator uses a mouse and joystick to tell the bot where to move its limbs. The robot processes the request, calculating the movement of each joint. These plan-into-motion equations are called inverse kinematics.

The DRC focuses on disaster response, but the teams envision much broader uses for their robots. The tasks here showcase attributes—dexterity, sensory awareness, and reliability—that robots will need to operate as our proxies in various environments. For example, a robotic attendant in a nursing home would have to open cabinets and doors that have different handles, latches, and heights, the same way it would when searching for survivors in a chemical-plant fire.

To have a future working for and with humans, limb control is crucial. "Robots are ridiculously strong, and things are fragile in the world," Daniel Lofaro, a graduate student at Drexel University and leader of the school's team, Hubo, says. "We need good feedback, and really quickly. The robot will turn the doorknob but tear the door off its hinges."

NE OF THE BELIEVERS IN THE FUTURE of mobile robotics is Google. It bought eight robotics businesses in 2013, a dramatic move that is reshaping the industry. Two of these newly acquired companies are represented at the DRC trials—Schaft, founded by researchers from the University of Tokyo, and Massachusetts-based Boston Dynamics. Google announced its purchase of Boston Dynamics just a week before the DRC, and the news brought fresh excitement and relevancy to the event.

Boston Dynamics employees at the DRC seem happy about their new bosses. "I've been a part of more than seven acquisitions during my career," one says. "This definitely doesn't feel like a bad one." Boston Dynamics made its money by fulfilling DARPA contracts for advanced prototypes, but employees say it's a good time to trade Pentagon funding for Google investment. In 2015, several high-profile DARPA programs are ending—as is Gill Pratt's tenure. "We weren't sure where the next millions were going to come from," the Boston Dynamics employee notes. "Now we do."

DARPA, despite its military trappings, is fairly open about sharing its marvelous new robots. Google is tightlipped about its advances. This proprietary mindset can already be seen at the speedway. The behavior of Team Schaft at the DRC stands in stark contrast to that of the other groups. Most teams bring along public relations staff to tout their universities and institutions, but Schaft is represented by one overworked Google employee from the California headquarters who didn't even know the DRC existed before she was ordered to Florida to support it. Schaft offers no access and no comment, and even chases off journalists standing in per-

LEFT: *HANDLERS FROM TEAM SCHAFT CHECK S-ONE'S POSITION, POWER CONNECTIONS, AND DIAGNOSTICS BEFORE THE ROBOT'S SUCCESSFUL DRIVE THROUGH A WINDING, 250-FOOT-LONG COURSE.*

RIGHT: *AS THE S-ONE ROBOT WIDENS ITS LEAD, IT ATTRACTS CROWD AND MEDIA ATTENTION—A THRONG FOLLOWS TEAM SCHAFT FROM EVENT TO EVENT, MUCH AS SPECTATORS FOLLOW THE LEADER DURING A PRO GOLF MATCH.*

mitted areas outside the garage, where signs establish a cordon and forbid any contact. Like a celebrity baby kept out of the limelight, the Schaft S-One robot draws a crowd whenever it appears. Other teams make sure to check out the cutting-edge machine–and indulge in some robot envy.

Christopher Rasmussen, a computer science professor at the University of Delaware and member of Team Hubo, aims his camera as S-One begins the terrain task. It must walk 40 feet over a jumble of increasingly uneven and steep concrete blocks.

Like all robots here, S-One is secured by a safety tether to protect it in case of a fall. The robot takes its first step, and the clock starts. S-One is stronger than most other competitors. Instead of using hydraulics, like Atlas, S-One is all electrical. Rather than relying on just a battery, it uses capacitors that can quickly supply lots of current to a limb. These millisecond bursts of power to its motors enable the robot to, for example, quickly generate torque in its knee to stabilize itself if it loses its balance.

Between each step, a cover in S-One's boxy body tilts open to reveal a laser radar. Rasmussen crouches to peer inside, trying to glean details. "The inside surface of the door could help reflect the laser to the ground," he guesses. He notes that the knees are constantly, minutely bending. "That makes it easier to balance," he says, "but it sort of looks like it's breathing."

S-One takes a step down from a block and scrapes a knuckle on concrete. "That would have knocked over a lesser robot," Rasmussen says. Unfazed, S-One continues down the pile and reaches level ground in less than 15 minutes. The cheers from observers in the stands cause teams at other events to turn their heads. Schaft scores three more points. The team hangs the unpowered robot on its wheeled carrier and spirits it off to the garage without a word to admirers.

By the middle of the second day, after S-One performs its final event, it has racked up 27 out of a possible 32 points, a first-place victory for the robot–and a clear win for Google.

The scoreboard tells dramatic tales as the trials come to an end. CHIMP, made by Carnegie Mellon's Team Tartan Rescue, is staging an epic comeback. One of the few robots that was designed just for the competition, it has 10,600 mostly homemade parts. CHIMP—a beefy machine encased in a red metal shell and with tracks on its arms for traveling on all fours—scores a perfect four points in two events by turning three valves and cutting a neat triangle in drywall. At day's end, it has scored 18 points, enough to place third.

Team Hubo is not so fortunate. Hampered by breakdowns, it finishes with just three points. "Disappointing," Lofaro says.

NASA has mixed results. JPL's RoboSimian places fifth with 14 points, but Valkyrie, made by the Johnson Space Center, tips over frequently and doesn't earn a single point.

Team WRECS and its Atlas robot, Warner, are on the edge. Warner had a strong second day at the DRC, driving the entire ATV course in just 6 minutes. No other team finished the drive as quickly, and no other Atlas drove across the finish line. Now it has to perform well in its last event—the walk through uneven terrain—to secure a coveted spot in the top eight.

The robot picks its way through the rubble, earning two points, then pauses at the top of the third brick pile, teetering as it tries to regain stability. Team WRECS calls this the Atlas dance. War-

Battle of Lifesaving Robots: Meet the Teams

Robot: Atlas
Teams: Seven. The challenge is open to software programmers, so DARPA gave each team its own Atlas.
Rundown: This 6.2-ft, 330-lb robot made by Boston Dynamics has 28 hydraulically actuated joints and a laser radar guidance package installed in its head.

Robot: RoboSimian
Team: NASA's Jet Propulsion Lab
Rundown: No one said the robots had to mimic humans exactly. NASA designed "an apelike disaster-recovery robot" with a squat profile that emphasizes stability over speed.

ner steadies itself, to cheers, but two steps into its descent, the robot topples and swings like a marionette in its safety harness. The crowd moans, then applauds. Two points is enough to propel WRECS into a tie for sixth place. As one of the top eight teams, they'll be back for the finals.

The DRC ends at sunset on Saturday. Hundreds of competitors file to the closing ceremonies, held at the speedway under a large white tent. They are exhausted, thrilled, disappointed, giddy, and happy to be here, surrounded by peers who understand what they are up to, and how hard it is to accomplish.

The emcee of the event introduces DARPA bigwigs, who give speeches. They proclaim what these believers already hold as canon—robots are coming. There's a sense among the crowd that this is the place and time that the mobilerobot revolution found its footing.

It won't happen again, not like this. Things can begin only once.

EDITOR'S NOTE

The DARPA Robotics Challenge finals are currently scheduled for June 5th, 2015 in Pomona, California. The winner will receive $2 million dollars, but it won't be easy. This time around, the humanoid robots will not be connected to any wires, including power cords or fall arrestors. Humans will not be able to intervene for any reason and all tasks must be completed within one hour (instead of four hours, as required during trials). —DHW

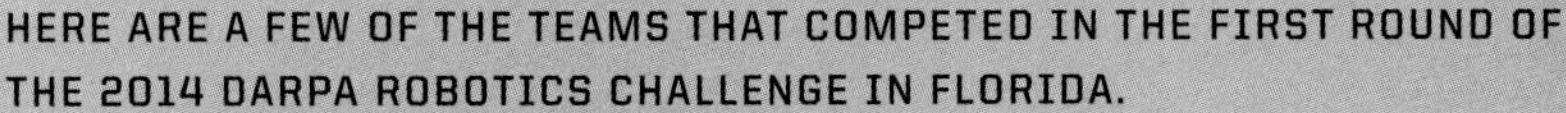

HERE ARE A FEW OF THE TEAMS THAT COMPETED IN THE FIRST ROUND OF THE 2014 DARPA ROBOTICS CHALLENGE IN FLORIDA.

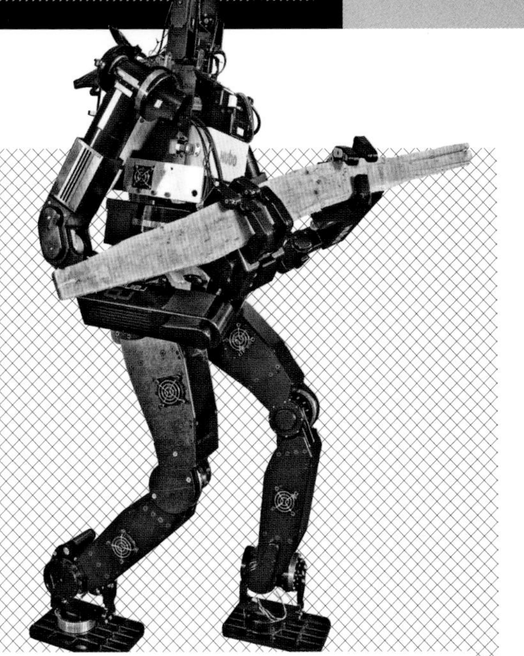

Robot: Hubo
Team: DRC-Hubo
Rundown: Designers compare this 4.7-ft, 120-lb android to a smartphone. Eight universities, with Drexel as lead, are each developing an application to control the robot as it does various tasks. "You want vehicle driving, load that app, and our robot drives the utility vehicle," the team's website reads.

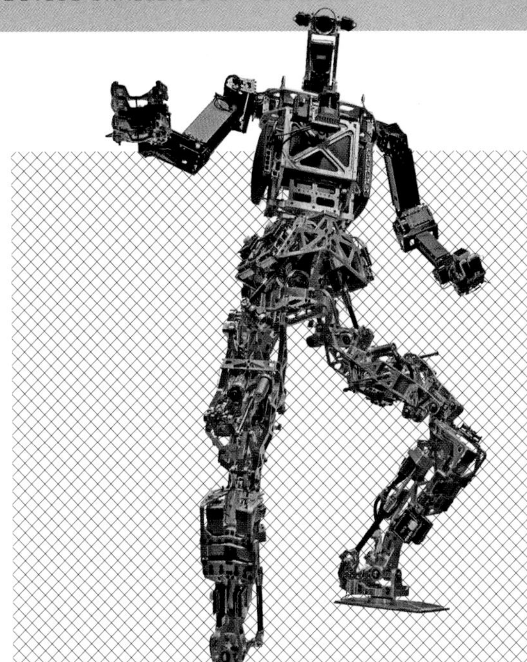

Robot: THOR (Tactical Hazardous Operations Robot)
Team: Virginia Tech
Rundown: Even the toughest robots need to be flexible. Elastic linear actuators drive this 5.6-ft robot's lower limbs, but it uses stiffer industrial-grade servos in its upper body.

WHEN CAN I LET GO OF THE WHEEL?

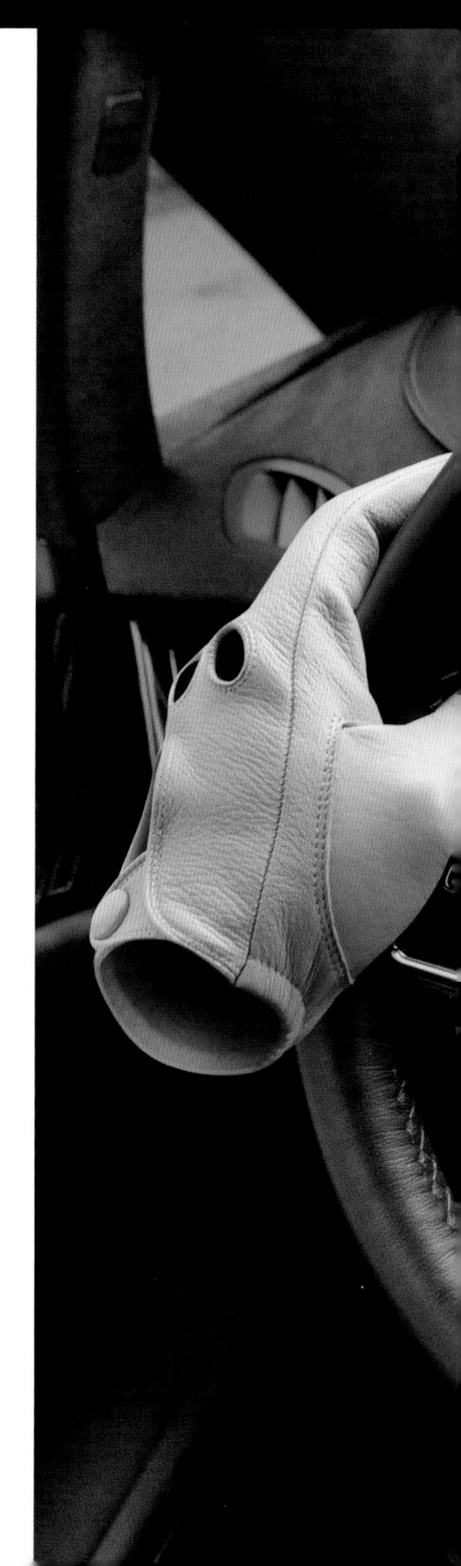

The self-driving car is coming.

BY **ANDREW DEL-COLLE**

This article was published in *Popular Mechanics* in 2013.

AS SOON AS I HIT THE SMALL BLACK LANE-centering button, I can feel the wheel stiffen in my hands. Suddenly, the light bar atop it begins glowing a soft green, signaling that the car has taken over the task of steering. I had already surrendered control of the acceleration and braking to the vehicle's adaptive cruise control, and now with my feet flat on the floor, I slowly unwrap my fingers and release the wheel. Sitting in the passenger seat is Jeremy Salinger, who works on General Motors' semiautonomous car program. He's seen the car do this dozens, possibly hundreds, of times, but, if only to reassure me, he states the obvious: "The car is in control now."

My remaining sense of caution keeps my hands nervously hovering above the wheel for a few seconds as our 2012 Cadillac SRX test vehicle hurtles forward at 60 mph, effortlessly bending into one of the sweeping turns of the oval track at GM's Milford Proving Grounds in Michigan. In the rear of the heavily modified SUV, an onboard computer collects data from an array of cameras and sensors to electronically read the road and determine the vehicle's position. Using this information, the car controls the electrically assisted steering to keep the car perfectly centered in the lane.

"Let's do something, like a lane change," Salinger says. The light bar switches from green to blue as I grab hold of the wheel and signal right, indicating that I'm overriding the system. Once we are settled into the new lane, the light bar changes back to green, and I let go of the wheel. We repeat the exercise, returning to the center lane of the five-lane track, and then my hands drop down to rest on my legs. Now fully at ease, I watch the wheel periodically turn and straighten as we speed around the track. The sight is both mesmerizing and unsettling–as if a ghost were driving. But this is no spooky apparition, this is GM's Super Cruise technology, and it could be available in a production vehicle within five years.

total power
0.5 kW
battery
PRNDL
engine
Classic
Select Layout
EV Range
32 mi
READY
Fuel Range
69 mi
POWER
RES
SET
SEL

Although GM hasn't confirmed a hard date for deployment of the system, it hopes to have Super Cruise as an option by the end of the decade, possibly as soon as the 2017 or 2018 model year. Super Cruise is part of a trend that has sneaked up on the car-buying public. The active safety systems in many modern production cars– forward-collision warning, park assist, adaptive cruise control, blind-spot monitoring, and others– are inching the vehicles we buy toward autonomy. By using a combination of sensors, radar, GPS, and cameras, these systems enable a car to interpret its surroundings and issue warnings or even brake the car if a collision with another vehicle or a pedestrian is imminent. As these systems become connected, our cars become smarter.

Technologically, GM's Super Cruise is similar to driver-assistance systems provided by Mercedes-Benz and other carmakers. But these other systems also require that the driver periodically touch the wheel to ensure he or she remains engaged in the driving experience. Where Super Cruise differs is that for the first time GM aims to allow the driver to completely let go of the wheel for extended periods of time. Functionally, this change is revolutionary revolutionary. By fully taking over the task of highway driving, Super Cruise might represent the most important shift in the driver-car relationship in automotive history.

SUPER CRUISE *IS A COMBINATION OF TWO TECHNOLOGIES. THE FIRST IS THE INCREASINGLY COMMON ADAPTIVE CRUISE CONTROL, WHICH USES A LONG-RANGE RADAR (MORE THAN 100 METERS) IN THE GRILLE TO KEEP THE CAR A UNIFORM DISTANCE BEHIND ANOTHER VEHICLE WHILE MAINTAINING A SET SPEED ❶. THE SECOND, LANE-CENTERING, USES MULTIPLE CAMERAS WITH MACHINE-VISION SOFTWARE TO READ ROAD LINES AND DETECT OBJECTS ❷. THIS INFORMATION IS SENT TO A COMPUTER THAT PROCESSES THE DATA AND ADJUSTS THE ELECTRICALLY ASSISTED STEERING TO KEEP THE VEHICLE CENTERED IN THE LANE. BECAUSE SUPER CRUISE IS INTENDED ONLY FOR HIGHWAYS, GM WILL USE THE VEHICLE'S GPS TO DETERMINE ITS LOCATION BEFORE ALLOWING THE DRIVER TO ENGAGE THE FEATURE ❸. IN ADDITION, GM IS ALSO CONSIDERING USING SHORT-RANGE RADARS (30 TO 50 METERS) ❹ AND EXTRA ULTRASONIC SENSORS (3 METERS) ❺ TO ENHANCE THE VEHICLE'S OVERALL AWARENESS. CARS WITH PARK-ASSIST SYSTEMS ALREADY HAVE FOUR SIMILAR SENSORS IN THE FRONT AND IN THE REAR OF THE CAR. GM IS ALSO EXPERIMENTING WITH COST-EFFECTIVE LIDAR UNITS, WHICH USE LASERS INSTEAD OF SOUND AND ARE MORE POWERFUL AND ACCURATE THAN ULTRASONIC SENSORS. IT'S UNCLEAR WHETHER LIDAR WILL MAKE IT INTO THE SAME VEHICLE AS SUPER CRUISE.*

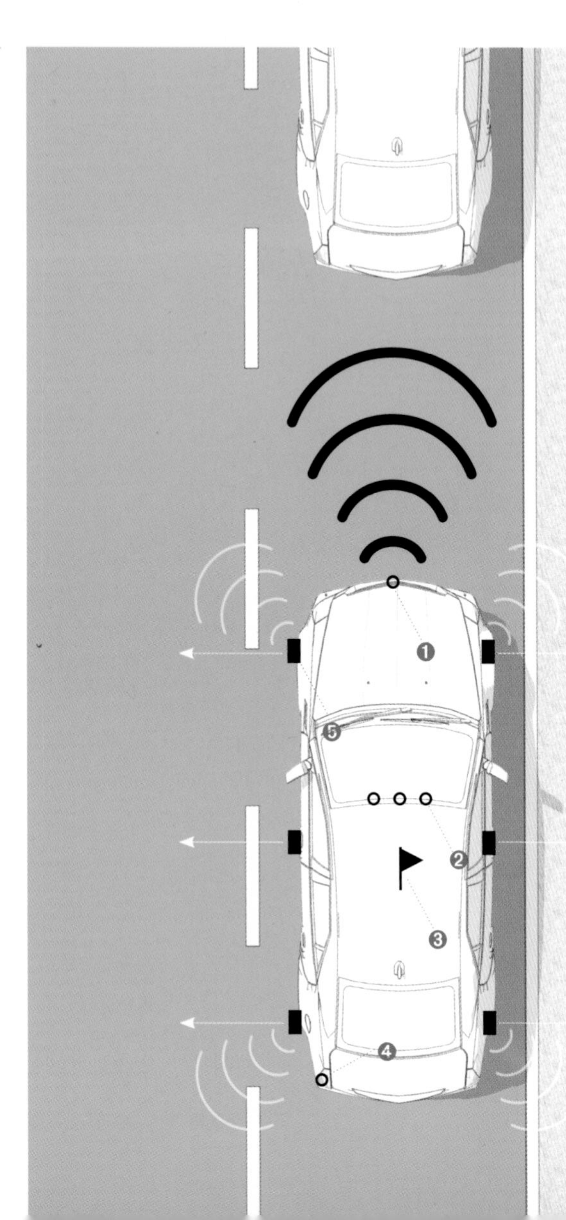

ET'S GET SOME SEMANTICS out of the way. The Super Cruise-equipped SRX I rode is a self- driving car, but it can't do everything on its own. Super Cruise will only be available under select highway conditions and will have speed and turning limitations, so think of it like autopilot for your car. By definition, this makes it semiautonomous (though much more capable Google's self-driving cars are also semiautonomous). A fully autonomous production car–where you plug your destination into the GPS and settle in for a 2-hour movie–is still a long way off, but the technology is advancing fast. Recently Audi and Volvo demonstrated vehicles that can park themselves; Volvo

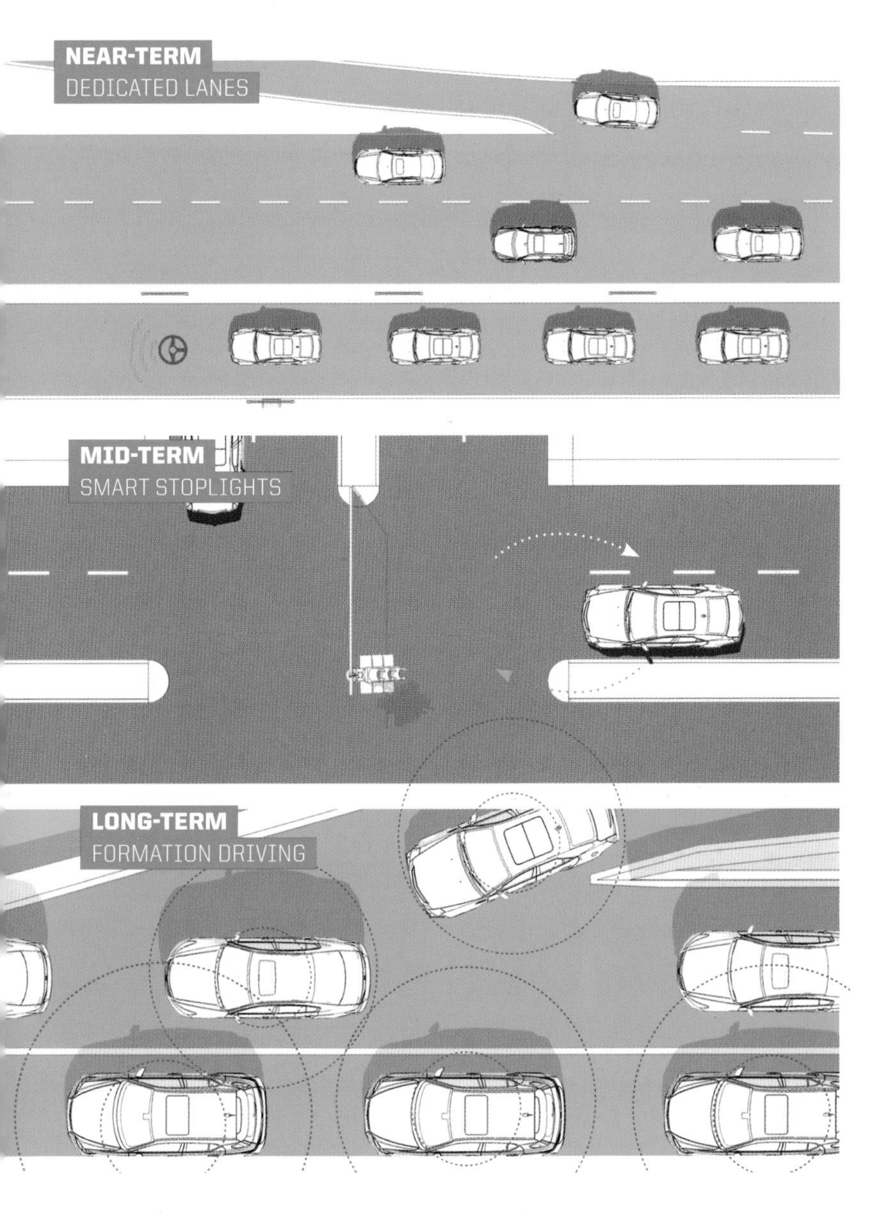

SEMIAUTONOMOUS *CARS AND VEHICLE-TO-VEHICLE COMMUNICATION COULD PROFOUNDLY CHANGE HOW TRAFFIC FLOWS AND EVEN HIGHWAY DESIGN. HERE ARE THREE LIKELY SCENARIOS.*

NEAR-TERM *ONCE ENOUGH CARS ON THE ROAD CAN DRIVE THEMSELVES, THE FIRST MAJOR CHANGE WE WILL PROBABLY SEE IS AN HOV-STYLE SELF-DRIVING LANE ON HIGHWAYS. JUST ENTER THE LANE, SET YOUR VEHICLE TO THE RECOMMENDED SPEED, AND LEAVE YOUR MONOTONOUS 3-HOUR HAUL TO THE CAR.*

MIDTERM *AS SEMIAUTONOMOUS VEHICLES ADVANCE, EXPECT THEM TO START COMMUNICATING WITH ONE ANOTHER AND EVENTUALLY EVEN WITH INFRASTRUCTURE. AT INTERSECTIONS, SMART STOPLIGHTS COULD AUTOMATICALLY TURN TO GREEN IF NO OTHER CARS ARE APPROACHING—WHICH WOULD SMOOTH OUT TRAFFIC FLOW, REDUCE IDLING, AND DECREASE ACCIDENTS.*

LONG-TERM *ONCE THE MAJORITY OF VEHICLES ON THE ROAD ARE SEMIAUTONOMOUS AND ABLE TO TALK WITH ONE ANOTHER, HIGH-SPEED MOTORING (WE'RE TALKING TRIPLE DIGITS HERE), FORMATION DRIVING, AND SEAMLESS MERGING COULD BE POSSIBLE.*

says this feature could be available as early as 2014. Still, systems such as Super Cruise challenge conventional notions of who or what is doing the driving, and this makes carmakers and regulators nervous. "I think the one thing we will always want to maintain, and the manufacturers will likely tell you the same thing, is that the driver's still responsible for the driving task," David Strickland, administrator of the National Highway Traffic Safety Administration, says. "That will never change." Super Cruise and other similar systems do more than just see the road. Using an array of sensors, lasers, radar, cameras, and GPS technology, they can actually analyze a car's surroundings.

According to Nady Boules, a director within GM's R&D, semiautonomous cars perform better than human drivers in many respects. Unlike people, they can continuously monitor all sides of the vehicle, can react almost instantaneously, and are impervious to distraction. They're also more efficient drivers. One recent paper presented at an Institute of Electrical and Electronics Engineers conference estimated that a fleet of cars with sensors could increase highway capacity by 43 percent. If the cars could also communicate with one another, that number jumps to an incredible 273 percent.

The NHTSA and several carmakers are already testing vehicles that can talk to each other. I got a chance to stop by GM's technical center in Warren, Mich., for a demonstration of this vehicle-to-vehicle (V2V) communication. In a large parking lot GM engineers had me run through a series of exercises with two cars, each equipped with GPS and a small Wi-Fi transponder. Using the GPS to determine its location and the transponder to talk with the other vehicle, my car warned me when the other vehicle braked hard in front of me, was in my blind spot, or was approaching from around a corner that had an obstructed view. Unlike with radar and cameras, line of sight isn't required for V2V communication, and the transponders are relatively inexpensive compared with the hardware required for most active safety systems. On its own, V2V technology is compelling, but once it is connected to the other active safety systems in a vehicle, it has powerful potential.

HIS IS LEGAL ONLY BECAUSE LAWS regulating self-driving cars didn't exist until recently. Currently, legislation is being considered on a state-by-state basis, with Nevada, California, and Florida having passed laws for the testing of self-driving cars on public roads; Michigan is expected to follow. Generally, these laws simply stipulate that someone must be in the car to take control of the car in case of an emergency. In May, the NHTSA released a set of suggested guidelines for the development of semiautonomous vehicles, but there are still no official federal regulations.

Bryant Walker Smith is a fellow at the Center for Internet and Society at Stanford Law School and the Center for Automotive Research. He points out that along with creating new laws, we will also need to review our existing laws as semiautonomous systems proliferate. "New York state, for example, requires drivers to keep one hand on the wheel at all times," Smith says. "Figure how that reconciles with Super Cruise." He goes on: "Regardless of what the traffic codes, what the vehicle codes specifically say, when a crash happens, how will judges and juries decide what behavior was negligent? So even if you're in a state that doesn't expressly say you have to have your hands on the wheel, will not having your hands on the wheel, will not paying attention, be considered negligent?"

This is a big issue and one that the NHTSA and carmakers are tiptoeing around by insisting that the driver will be responsible at all times. Nevertheless, semiautonomous technologies have huge implications for insurance agencies. Loretta L. Worters is the vice president of communications for the Insurance Information Institute. According to her, systems that improve safety are welcomed by the insurance industry, but as our cars turn into semiautonomous vehicles, we are entering new territory. "For example, it would also likely increase the potential liability of the manufacturer and maintenance/repair business if an accident could be traced to a design or maintenance failure," Worters says.

Considering that semiautonomous cars rely on a network of active safety systems to work, they're pretty safe. The Insurance Institute for Highway Safety has found that we are already seeing the benefits of systems such as forward-collision warning in the reduction of accidents. That said, they're not 100 percent fail-proof. Chris Urmson leads Google's self-driving car program and is an adjunct professor at Carnegie Mellon University, where he also worked on DARPA's autonomous car challenges. Although Google's test fleet has logged more than 500,000 miles without a major mishap, Urmson admits that the technology will never be flawless. "I think we can make vehicles that are better than human-driven vehicles, but something that won't fail is just an impossibility," he says.

Aside from a possible system failure, semiautonomous technologies also have general limitations. My Super Cruise experience was actually my second attempt at seeing the system in action. On my first trip to the proving grounds, a snowstorm thwarted my mission. The SRX's cameras were unable to read the lines on the snow-covered track, leaving Super Cruise ineffective. That's no small hindrance.

It's important to note that no matter how capable the vehicle might be, a semiautonomous car cannot be left unattended—no napping behind the wheel. And the chance of someone falling asleep is a serious concern for carmakers. Eric Raphael also works on Super Cruise and gave me a technical walk-around of the modified SRX on my visit. "We can't let someone crawl in the back seat and take a nap, but we know they might try that," Raphael says. To prevent a driver from spacing out or worse, GM has developed a monitoring system that will use audial, visual, and possibly haptic feedback to prompt the driver to take over if it detects an emergency or that the driver is too distracted. For example, the light bar at the top of the steering wheel turns bright red when the driver needs to take control of the car.

Of course, at 60 mph and in the middle of heavy traffic, the system can't physically force the driver to reengage—and that's more than a little worrisome. Bryan Reimer is a research scientist at MIT's AgeLab and the associate director of the New England University Transportation Center. "Man, who's saying you're going to be awake when you actually need to put your hands back on there?" Reimer says. "All the literature in psychology tells us that we are terrible overseers of highly autonomous systems. We fall asleep. We get bored. We're inattentive."

Generally speaking, the more we use autonomous systems, the more we come to rely on them. So it stands to reason that the less we drive, the rustier we'll get at it. Citing examples such as the pilot error responsible for the crash of Air France 447, Reimer stresses the need for caution as we integrate these technologies into our cars: "If we look at other domains, and I think the aviation world is a great example of this, it is now a well-accepted fact that autopilot is decreasing pilot skill."

Most forward-collision warning systems today can identify another car, a pedestrian, an animal, or a bike, and some can even activate the brakes in low-speed to midspeed situations. But as for avoiding an obstacle, no system out there does that yet. Salinger says cars in the future could have this ability, though it's not something GM is currently developing. Even the possibility raises an important issue: If a split-second steering decision needs to be made, would you want to leave it up to your car? Say the decision comes down to hitting a child in the road or swerving and potentially hitting a telephone pole. Would the car put passengers in danger to avoid hitting the kid? A technological issue swiftly becomes a moral one.

And the most important point of all: As long as there is a traditional steering wheel in your car, you'll be able to drive. Now, things will get interesting if these semiautonomous systems become so safe and reliable that a human driving is actually viewed as irresponsible by insurers. That won't happen any time soon, but eventually it could. In one exercise during my Super Cruise demonstration, a Chevy Volt darted in front of our SRX as the car was driving itself. Still traveling at 60 mph, the SRX immediately braked, adjusted its speed to maintain a safe distance behind the Volt, and slightly shifted the wheel to stay centered in the lane. In a situation where a person might overreact and jerk the wheel, causing the car to careen into another vehicle or off the road, I didn't do a thing. I just sat there. I don't think I even flinched. I love driving, but who wouldn't want that as an option in their next car?

EDITOR'S NOTE

Google has been busy making self-driving cars a reality. After hiring Chris Urmson (who is still adjunct faculty at Carnegie Mellon University), Google introduced its first prototype self-driving car via the Google Official Blog on May 27th, 2014. The golf cart-sized vehicle has room for two passengers, padded bumpers, and no steering wheel. —DHW

World's Smartest Lamp Post

MICHIGAN-BASED ILLUMINATING CONCEPTS IS SELLING WIRED LIGHT POSTS THAT DO MORE THAN BRIGHTEN A STREET CORNER—THEY FORM AN INFORMATION NETWORK.

Mesh Transceiver
Transmits and receives data from a central receiver and poles, forming a street-level network.

Smart-Grid Lighting
Provides on-demand light levels; batteries keep post active if disconnected from the grid.

Digital Street Sign
Can be used for advertising, traffic direction, or preprogrammed emergency guidance based on sensor data.

Speaker
Delivers music or communicates alerts across the network or to a specific post.

Digital Signage
Can change messages to reflect planned or unplanned events.

Emergency Light Button Sensors
Can detect chemical, biological, or nuclear materials in the air, while seismic monitors warn of earthquakes.

Water Detection
Real-time data on water levels; useful in flood prone areas.

Intellistreets System
Ordered by Philadelphia and Chicago.

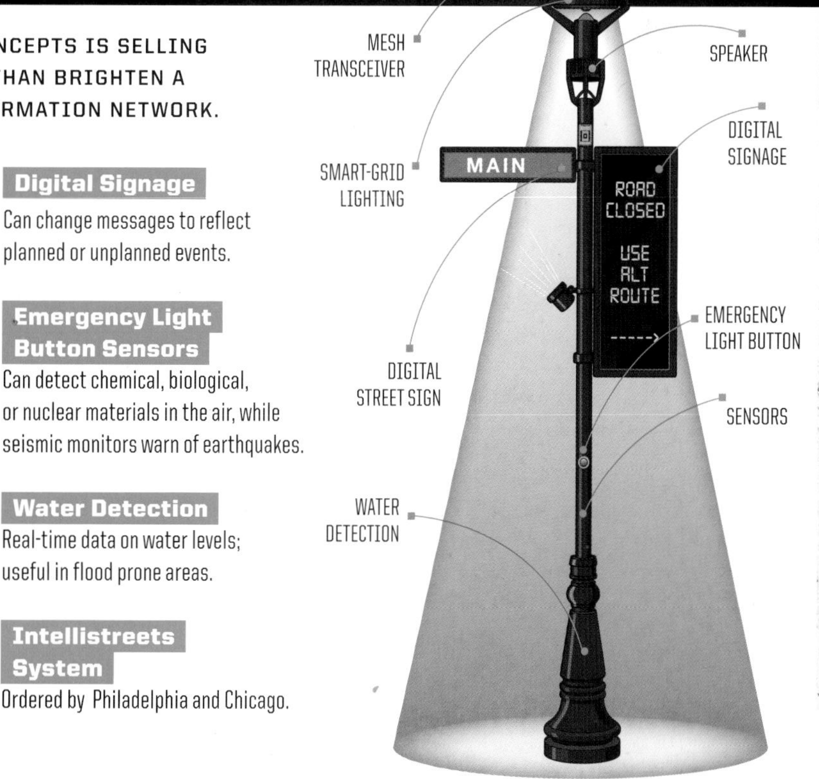

Tech Terms: The Concepts You Need to Know

Cognitive radio

The airways are getting crowded, thanks to smartphone and tablet data transmission that doubles every year. One solution: cognitive radio devices, whose signals automatically jump back and forth between frequencies in a fraction of a second to find open spectrum. A prototype developed at Rutgers University can switch to a new frequency in less than 50 microseconds, taking advantage of openings on the AM and FM radio, TV, and cellular frequency bands, while sending eight times the data of a typical home wireless system. And Florida-based xG Technology has already set up a demo network in Fort Lauderdale that uses cognitive radio for mobile broadband and VoIP links. Crucially, the FCC announced in September a pending rule change that will allow spectrum-sharing technologies, such as cognitive radio, to use previously restricted frequencies.

Single-board computer

With credit-card size computers, made of a single circuit board, the Raspberry Pi Foundation aims to take science and technology to schools around the world. The Raspberry Pi computer—generically known as a singleboard computer—runs on Linux and can be plugged directly into a TV or keyboard to play HD video and run software for word processing and games. To keep down costs, the unit has no Flash or hard drives and, instead, relies on SD cards for system information and storage.

Plug-and-play satellites

CubeSats are tiny satellites 4 inches long and less than 3 pounds each. They're easy enough to build that amateurs and high school students have done it—and they've helped spark a total rethinking of what it takes to send a satellite into space. Researchers at the Air Force Research Laboratory and elsewhere have been working on a new plug-and-play approach that can be applied to satellites up to 1,000 pounds. By standardizing the usual components and developing a common language for the parts to communicate with one another, designers can avoid rethinking the gyroscope for every new project. The result: A satellite can be designed and built in six days instead of six years. Northrop Grumman has adopted this approach for its Modular Space Vehicles (MSV), which will allow military commanders to order custom tactical satellites and receive them within weeks. The first MSV is expected to launch in 2013.

Inductive EV charging

It's not terribly complicated: An electric current in one coil of wire generates an electromagnetic field, and that induces a voltage in another nearby coil of wire. Presto, you've just charged your battery with no wires! Not terribly new either. (Remember the Palm Pre's optional inductive charger back in 2009? Exactly.) Infiniti plans to release a model that charges from a coil embedded in the ground under your parking spot. The technology is also showing up in consumer electronics, including Nokia's new Lumia phones. The critical question: Will the Wireless Power Consortium, whose aim is to impose standards so any wireless charger can work with any device, get companies to cooperate, so that everyone's parking spots and portable devices can play nice?

Co-robotics

Old-school industrial robots work best alone—try to help an assembly-line welding bot, and you'll probably get welded. But the next generation of robots will work closely with humans, augmenting our capabilities and compensating for our weaknesses. That's why the National Robotics Initiative is pouring up to $50 million a year into co-robotics. The initiative is backed by agencies ranging from NASA (robots to help astronauts and explore terrain where humans can't go) and the National Institutes of Health (robot surgery for everyone and home care for the elderly) to the Department of Agriculture (robots that can deworm animals and sense fruit ripeness). A key first step to robot-human interaction: full-size humanoids such as UPenn and Virginia Tech's SAFFiR (right), which will help fight fires.

Space fence

Even the final frontier is getting crowded these days, as illustrated by the 2009 smashup of an American communications satellite with a Russian one. The U.S. is currently tracking space objects with an outmoded system commissioned in 1961. But that will start to change when crews on Kwajalein Island in the northern Pacific start construction of the first radar site in the new $3.5 billion Space Fence network. The system uses high-frequency radar to detect objects as small as a softball at a distance of 1,200 miles and can perform "uncued tracking," which means it can track objects it hasn't already registered. The result: Once it's up and running in 2017, the Space Fence will keep tabs on more than 200,000 objects in low- and medium-Earth orbit.

THE *NEXT GENERATION OF ROBOTS WILL WORK CLOSELY WITH HUMANS.*

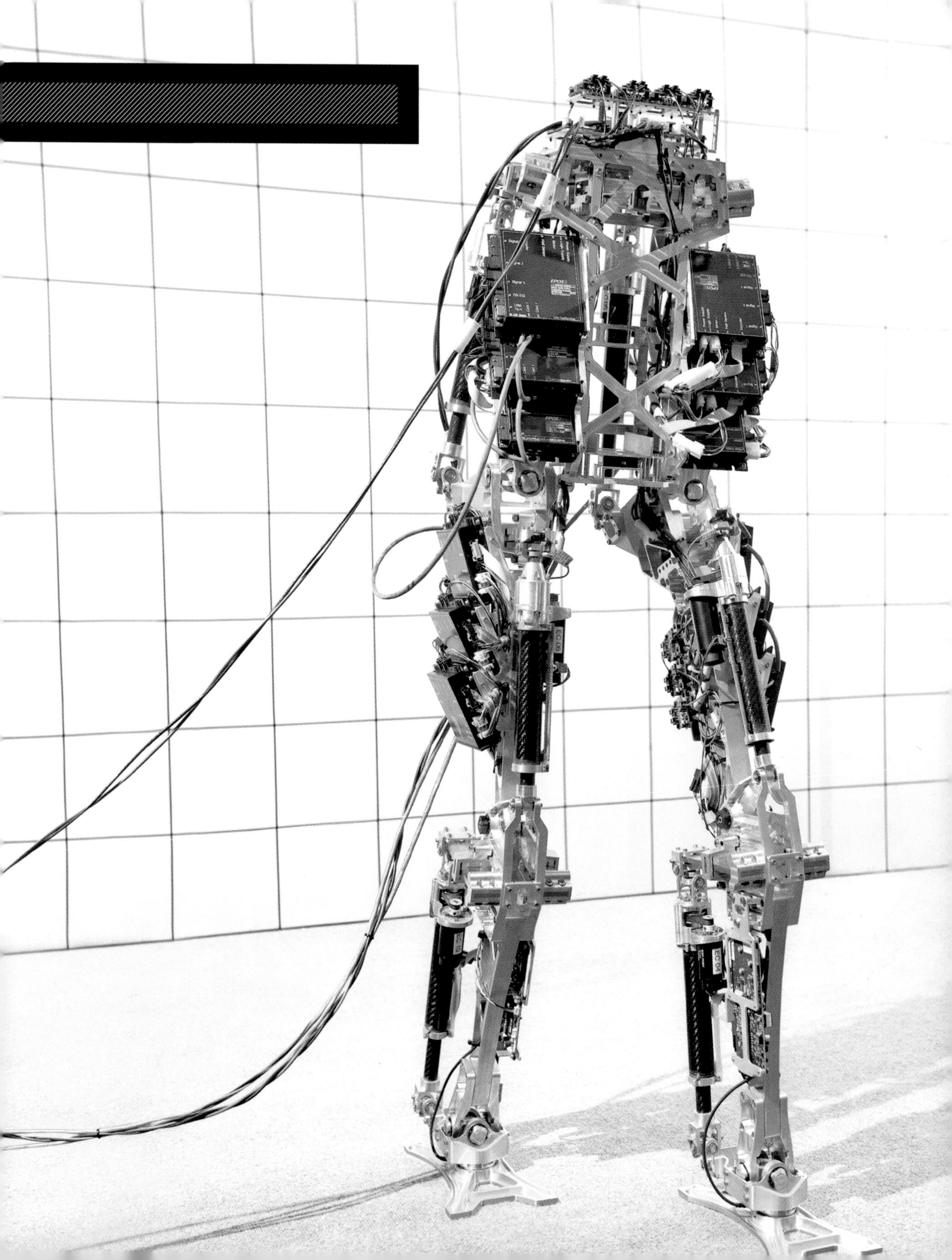
EPOS
EC 08
EC 10
EC 04

→ MABEL: Teaching Robots to Walk

Walking, that fundamental human activity, seems simple: Take one foot, put it in front of the other; repeat. But to scientists, bipedalism is still largely a mystery, involving a symphony of sensory input (from legs, eyes, and inner ear), voluntary and involuntary neural commands, and the synchronized pumping of muscles hinged by tendons to a frame that must balance in an upright position. That makes building a robot that can stand up and walk in a world built for humans deeply difficult.

But it's not impossible. Robots such as Honda's Asimo have been shuffling along on two feet for more than a decade, but the slow, clumsy performance of these machines is a far cry from the human gait. Jessy Grizzle of the University of Michigan, Ann Arbor, and Jonathan Hurst of Oregon State University have created a better bot, a 150-pound two-legged automaton named MABEL that can walk with a surprisingly human dexterity. MABEL is built to walk blindly (without the aid of laser scanners or other optical technologies) and fast (it can run a 9-minute mile). To navigate its environment, MABEL uses contact switches on its "feet" that send sensory feedback to a computer. "When MABEL steps off an 8-inch ledge, as soon as its foot touches the floor, the robot can calculate more quickly and more accurately than a human the exact position of its body," explains Grizzle. MABEL uses passive dynamics to walk efficiently—storing and releasing energy in fiberglass springs—rather than fighting its own momentum with its electric motors.

The quest for walking robots is not purely academic. The 2011 Fukushima Daiichi nuclear disaster highlighted the need for machines that could operate in hazardous, unpredictable environments that would stop wheeled and even tracked vehicles. Grizzle and Hurst are already working on MABEL's successor, a lighter, faster model named ATRIAS. But there's still plenty of engineering to be done before walking robots can be usefully deployed, walking into danger zones with balance and haste but no fear.

ONE *SMALL STEP FOR ROBOTS JESSY GRIZZLE (LEFT) DESIGNED THE SOFTWARE THAT HELPS BIPEDAL ROBOT MABEL NAVIGATE UNFAMILIAR TERRAIN AT UP TO 6.8 MPH. JONATHAN HURST (RIGHT) BUILT THE ROBOT'S LIGHTWEIGHT FRAME, DRAWING INSPIRATION FROM THE GAIT OF RUNNING BIRDS.*

MICHIGAN
M
Carnegie Mellon

WORLD OF TOMORROW

➡ Make way for digital scrolls, robotic trainers, and self-repairing bridges. It's never easy to predict the future, but we've decided to give it a try (with a little help from some very smart friends).

BY **THE EDITORS OF *POPULAR MECHANICS***

This article was published in *Popular Mechanics* in 2012.

IN 1949, FORECASTING THE RELENTLESS march of science, *Popular Mechanics* said, "Computers in the future may weigh no more than 1.5 tons."

Of course, what the article did not anticipate were two of the most pivotal inventions in human history: the transistor, which came into widespread use in the mid-1950s, and the integrated circuit, or microchip, which intensified the march toward miniaturization a decade later. The first fully transistorized computer, the IBM 608, hit the market in late 1957. It weighed 1.2 tons.

And therein lies the dilemma for anyone who dares speculate about the future. Even if you're accurate in the near term—and *Popular Mechanics'* 1949 story was a respectable piece of prognostication—a prediction can look naive or ridiculous when viewed a generation or two later. But we don't let that stop us.

As a magazine about science and engineering, it's our job to explain how technologies just now coming into view might shape the future. We turned to scores of experts—scientists, engineers, and many longtime *Popular Mechanics* contributors and consultants—to help us sketch the rough shape of the next century. We canvassed our experts about the nature of future changes and when key breakthroughs might occur. How did we do? There's only one way to know for sure: Check back with us in 110 years.

DRUGS *WILL BE TESTED ON "ORGAN CHIPS" THAT MIMIC THE HUMAN BODY. NOW UNDERGOING TRIALS IN 15 RESEARCH INSTITUTIONS, THE SILICON CHIPS HOUSE LIVING KIDNEY OR LUNG CELLS, AND SIMULATE BLOOD AND OXYGEN FLOW TO MIRROR THE ACTIONS OF REAL ORGANS, REDUCING THE NEED FOR ANIMAL TESTING AND SPEEDING UP DRUG DEVELOPMENT. IN THE MIDST OF A PANDEMIC, THAT WOULD BE CRUCIAL.*

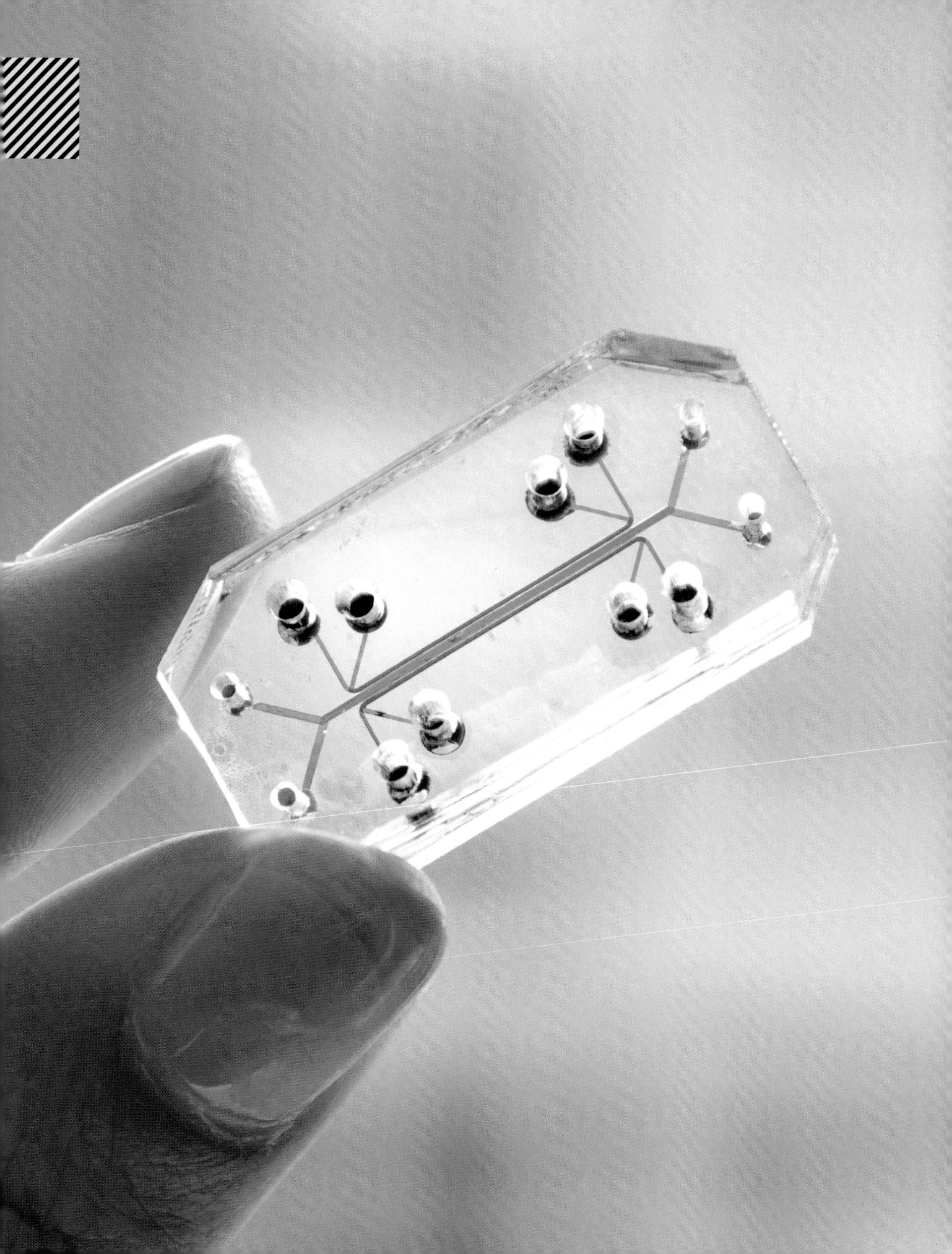

Your car will be truly connected. It will communicate with traffic lights to improve traffic flow. It will interact with other vehicles to prevent accidents. It will let you drag and drop a playlist from your home network. It will find the gas station with the deepest discount and handle the payment. It will notify you when someone dents your door and supply footage of the incident.

HAT CAR PART YOU NEED WILL BE SCULPTED inside a 3D printer. Dentists are already using this modern tech wonder to transform laser scans of your mouth into custom-fit appliances for your teeth. But that's a fraction of what the machine can do. When a 3D printer costs the same as, say, an HDTV, you will use one of your own to download all sorts of useful things, marveling as it creates each item layer by layer from plastic, rubber, titanium—you name it. Just imagine your future self printing a birthday cake, a Rolex, or a catalytic converter for the car. In time, you'll even be able to download prescription medicine.

Passwords will be obsolete. IBM says it will happen in five years. Who are we to disagree? Apple and Google are designing face-recognition software for cellphones. DARPA is researching the dynamics of keystrokes. Others are looking into retinal scans, voiceprints, and heartbeats. The big question, it seems, is what will you do with all that time you used to spend dreaming up new ways to say JZRulz24/7!

→ 2012-2022

ELECTRIC CARS WILL ROAM (SOME) HIGHWAYS. Who says you can't road-trip in a Tesla? In a few years, the 1,350-mile stretch of Interstate 5 spanning Washington, Oregon, and California will be lined with fast-charging stations—each no more than 60 miles apart. In some areas you will find stations to the east and west too. Don't get any bright ideas, though. If you try to cross the country, you won't get much farther than Tucson.

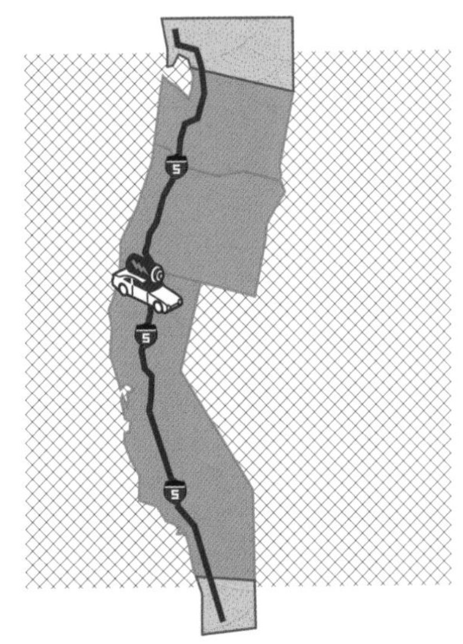

PEOPLE WILL BE FLUENT IN EVERY LANGUAGE. With DARPA and Google racing to perfect instant translation, it won't be long until your cellphone speaks Swahili on your behalf.

SOFTWARE WILL PREDICT TRAFFIC JAMS BEFORE THEY OCCUR. Using archived data, roadside sensors, and GPS, IBM has come up with a modeling program that anticipates bumper-to-bumper congestion a full hour before it begins. Better yet, the idea proved successful in early tests—even on the Jersey Turnpike.

NANOPARTICLES WILL MAKE CHEMOTHERAPY FAR MORE EFFECTIVE. By delivering tiny doses of cisplatin and docetaxel right to cancerous cells, the mini messengers will significantly reduce the pain and side effects of today's treatments.

ATHLETES WILL EMPLOY ROBOTIC TRAINERS. Picture a rotor-propelled drone that tracks a pattern on your T-shirt with an onboard camera. Now imagine it flying in front of you at world-record pace. That's just the start—a simple concept developed by researchers in Australia.

BRIDGES WILL REPAIR THEMSELVES WITH SELF-HEALING CONCRETE. Invented by University of Michigan engineer Victor Li, the new composite is laced with microfibers that bend without breaking. Hairline fractures mend themselves within days when calcium ions in the mix react with rainwater and carbon dioxide to create a calcium carbonate patch.

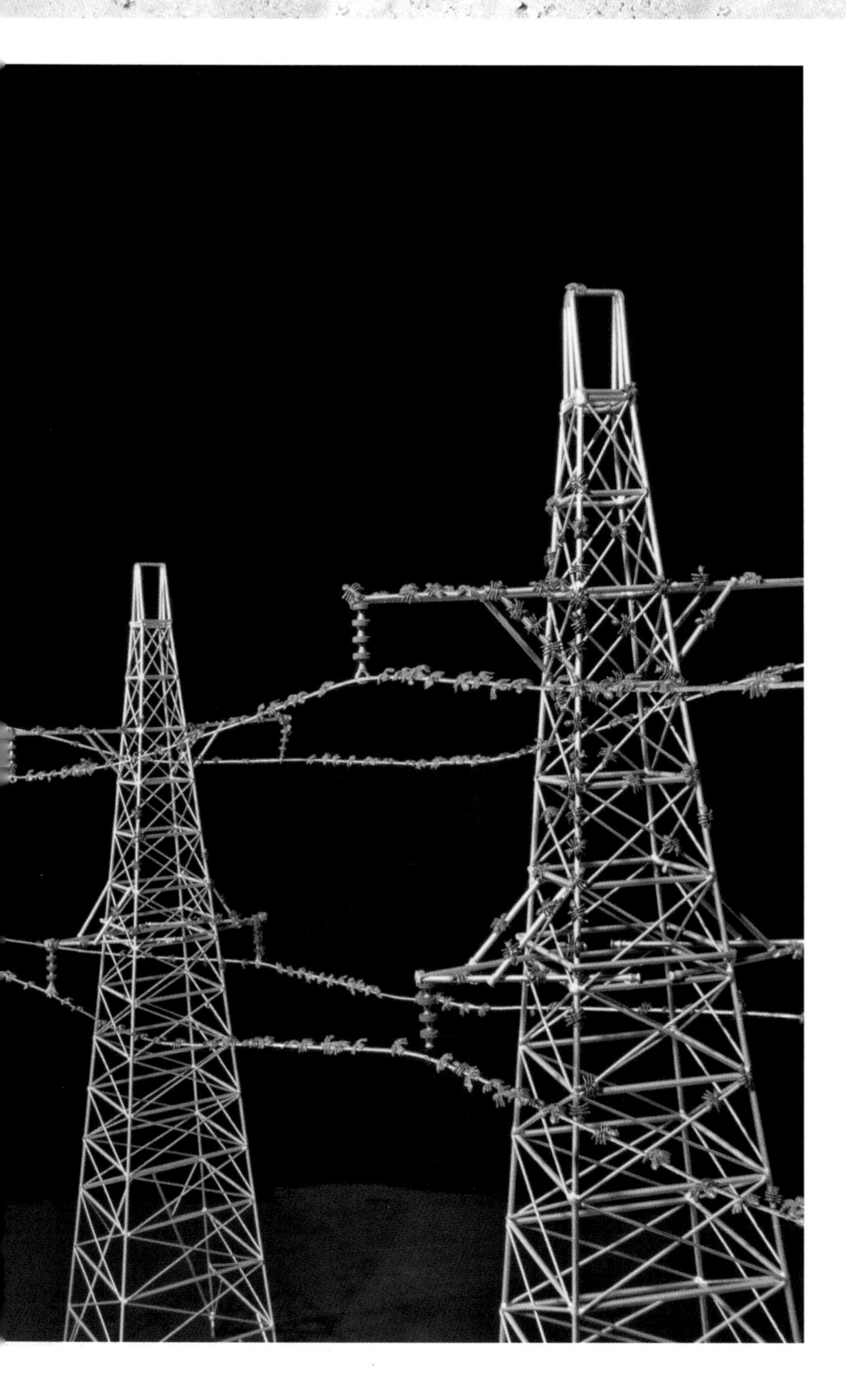

DIGITAL *"ANTS" WILL PROTECT THE U.S. POWER GRID FROM CYBER ATTACKS. PROGRAMMED TO WANDER NETWORKS IN SEARCH OF THREATS, THE HIGH-TECH SLEUTHS IN THIS SOFTWARE, DEVELOPED BY WAKE FOREST UNIVERSITY SECURITY EXPERT ERRIN FULP, LEAVE BEHIND A DIGITAL TRAIL MODELED AFTER THE SCENT STREAMS OF THEIR REAL-LIFE COUSINS. WHEN AN ANT DESIGNED TO PERFORM A TASK SPOTS A PROBLEM, OTHERS RUSH TO THE LOCATION TO DO THEIR OWN ANALYSIS. IF OPERATORS SEE A SWARM, THEY KNOW THERE'S TROUBLE.*

→ Eye Spy

Manufacturers of eye scanners that identify people by their irises tout the technology as the 21st-century equivalent of fingerprint analysis. But scanners have been limited by their 6-inch range, as well as their sensitivity to movement and obstructions (such as stray lashes). Now Honeywell has built a Combined Face and Iris Recognition System (CFAIRS), which extends the range of iris scans to 16 feet. CFAIRS shoots a high-resolution video image of the iris, then cross references it with biometric databases. "The software flattens the iris, unfolding it into a two-dimensional speckle pattern that looks like a standard bar code," says Dan Sheflin, Honeywell's vice president of advanced technology for automation and control solutions. At an airport, the 2.5-foot-tall machine would pan and tilt 120 degrees to survey travelers filing into customs. "This looks through masks and glasses, scans at off-angles, and captures people who are moving," Sheflin says. "You can be walking down a corridor and not even know it's being done." Local and federal law enforcement agencies are creating standards for collecting biometric data. For now, the tool's utility is limited by the depth of the databases.

Self-cleaning buildings will help us fight smog. When sunlight strikes their aluminum skin, a titanium dioxide coating releases free radicals, which break down the grime and convert toxic nitrogen oxide molecules in the air into a harmless nitrate. Everything washes away in the rain.

Your clothes will clean themselves too. Engineers in China have developed a titanium dioxide coating that helps cotton shed stains and eliminate odor-producing bacteria. To revive your lucky shirt after a night of poker, you need only step into the sun.

Data will be measured in zettabytes. According to the International Data Corporation, the volume of digital content created on the planet in 2010 exceeded a zettabyte for the first time in history. By the end of this year, the annual figure will have reached 2.7 zettabytes. What exactly does a zettabyte look like? Well, if each byte were a grain of sand, the sum total would allow you to build 400 Hoover Dams.

Smart homes will itemize electric, water, and gas bills by fixture and appliance. Shwetak Patel, a 30-year-old MacArthur Fellow, is working on low-cost sensors that monitor electrical variations in power lines to detect each appliance's signature. He has already used pressure changes to do the same for gas lines and water pipes. It's up to you to pinpoint where the savings lie.

EDITOR'S NOTE
A United States patent for CFAIRS was granted in April of 2014. As of publication, the product is not in active use in airports— yet. —DHW

The Brain Trust Says:

Within 20 years

Self-driving cars will hit the mainstream market. Battles will be waged without direct human participation (think robots or unmanned aerial vehicles).

Within 30 years

All-purpose robots will help us with household chores. Space travel will become as affordable as a round-the-world plane ticket.

Within 50 years

We will have a colony on Mars.

SCROLLS *WILL REPLACE TABLETS. RESEARCHERS HAVE ALREADY REPRODUCED WORDS AND IMAGES ON THIN PLASTIC DIGITAL DISPLAYS. IF THEY WANT THOSE DISPLAYS TO COMPETE WITH THE IPAD, THEY NEED TO FINE-TUNE THE COLOR AND REFINE THE SCREENS SO YOU CAN PUT YOUR FEET UP AND WATCH LEBRON THROW DOWN ON YOUTUBE.*

DRONES *WILL PROTECT ENDANGERED SPECIES. GUARDING AT-RISK ANIMALS FROM POACHERS WITH FOOT PATROLS IS EXPENSIVE AND DANGEROUS. THIS SUMMER RANGERS IN NEPAL'S CHITWAN NATIONAL PARK PREVIEWED A SAVVY SOLUTION: HAND-LAUNCHED DRONES ARMED WITH CAMERAS AND GPS PROVIDED AERIAL SURVEILLANCE OF THREATENED INDIAN RHINOS.*

RESCUERS *WILL USE ELECTRONIC NOSES TO LOCATE DISASTER VICTIMS. SOME DEVICES WILL USE AN ARRAY OF SENSORS TO RAPIDLY DETECT CARBON DIOXIDE, AMMONIA, AND ACETONE RELEASED INTO THE RUBBLE VIA BREATH, SWEAT, AND SKIN. OTHERS SNIFF OUT CHEMICAL COMPOUNDS FROM HUMAN REMAINS BURIED 3 FEET UNDERGROUND. ALL KEEP WORKING LONG AFTER THE DOGS HAVE RETIRED TO THEIR KENNELS.*

USAF

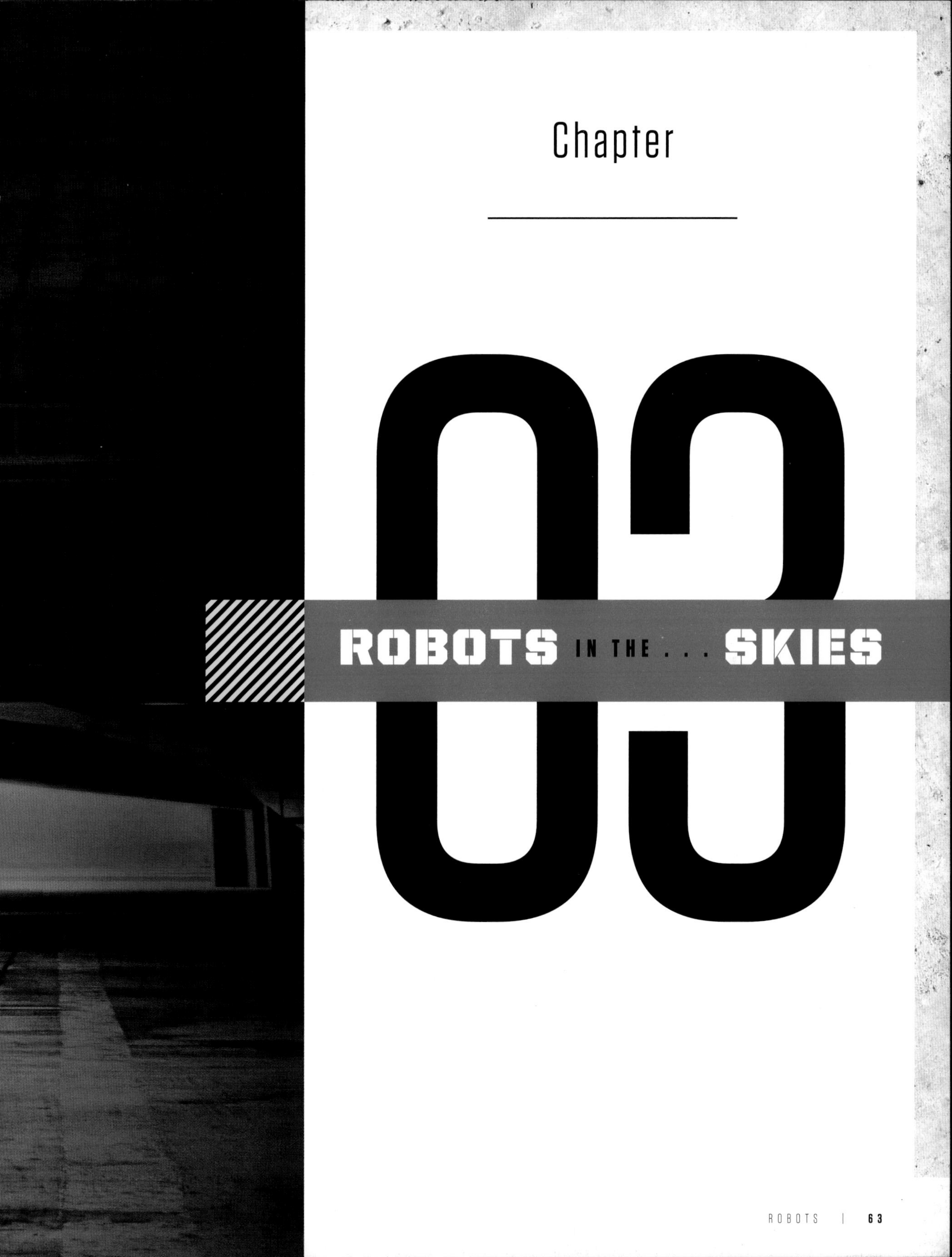

Chapter

03

ROBOTS IN THE . . . SKIES

DRONE SKIES

The proliferation of unmanned aircraft is already outpacing the regulations that govern them. Is U.S. airspace big enough for a deluge of drones?

BY **RICHARD WHITTLE**

(ADDITIONAL REPORTING BY GLENN DERENE)

This article was published in *Popular Mechanics* in 2013.

T'S A QUIET MORNING IN SAN FRANCISCO, with soft sunlight illuminating patches of thick fog billowing over the Golden Gate Bridge. A solitary unmanned aircraft–a 4-pound, battery-powered wedge of impact-resistant foam with a 54-inch wingspan, a single pusher-propeller in the rear, and a GoPro video camera attached to its body–quietly approaches the landmark.

Raphael "Trappy" Pirker controls the aircraft from a nearby hill. The bridge is within sight, but the 29-year-old enjoys the scenery through virtual-reality goggles strapped to his head. The drone's eye view is broadcast to the goggles, giving Pirker a streaming image of the bridge that grows larger as he guides the radio-controlled aircraft closer.

Pirker, a multilingual Austrian and a master's student at the University of Zurich, is a cofounder of a group of radio-control-aircraft enthusiasts and parts salesmen called Team BlackSheep. This California flight is the last stop of the international group's U.S. tour. Highlights included flights over the Hoover Dam, in Monument Valley, down the Las Vegas Strip, and through the Grand Canyon. The team has also flown above Rio de Janeiro, Amsterdam, Bangkok, Berlin, London, and Istanbul.

The Golden Gate Bridge now fills the view inside Pirker's goggles. He's not a licensed pilot, but his command over the radio-controlled (RC) aircraft is truly impressive. The drone climbs to the top of the bridge, zips through gaps in the towers, dives toward the water, and cruises along the underside of the bridge deck. Months later, the

RAPHAEL *PIRKER, SHOWN OPERATING A QUADROTOR IN HONG KONG, CONDUCTS DAREDEVIL FLIGHTS WITH CAMERA-EQUIPPED DRONES NEAR GLOBAL LANDMARKS, TO THE DELIGHT OF INTERNET VIEWERS AND THE DISMAY OF MANY GOVERNMENTS. HIS COMPANY, TEAM BLACKSHEEP, USES THE VIDEOS TO SELL DRONE EQUIPMENT.*

→ Robotics Aquatics

EIGHT ROBOTIC ARMS UNDER THE SKIN CONTROL MOVEMENT.

A RECHARGEABLE NICKEL-METAL HYDRIDE BATTERY POWERS THE ROBOT.

A THICK, FLEXIBLE SILICONE COVER MIMICS THE UMBRELLA OF A JELLYFISH.

DESCRIPTION: Drones aren't just proliferating in the sky—they'll soon populate the seas, where engineers are drawing inspiration from nature's own designs. Dispatched by governments and researchers, these unmanned craft will patrol harbors, track wildlife migration, and investigate shipwrecks.

1. Cyro

Engineers at Virginia Tech, funded by the U.S. Navy, developed Cyro to be an endurance undersea-monitoring machine. The 170-pound, silicone-skinned robotic jellyfish can operate in the ocean for months. The university researchers are considering how the Cyro could help clean up oil spills, but the Navy might use the technology for subtle underwater surveillance.

ACM-R5H

This amphibious snake comes with a glowing, cyclopean camera head and ridges—tiny wheels and paddles for motion—that line its segmented body. The system, developed by Japanese company HiBot in conjunction with the Tokyo Institute of Technology, is designed for search-and-rescue operations that would require wriggling through gaps too narrow for human first responders.

3. Crabster CR200

Developed at the Korea Institute of Ocean Science & Technology, the Crabster is a 6½-foot-tall 1322-pound tethered robot that next July will be used to inspect shipwrecks and track marine life. The giant bot is armed with LED spotlights, sonar, cameras, and navigation sensors, and its legs are powerful enough to stay on course, even when going against the current .

self-described RC Daredevils post the footage on YouTube, where nearly 60,000 viewers watch it.

Team BlackSheep is willfully—gleefully, really—flying through loopholes in the regulation of American airspace. The Federal Aviation Administration (FAA) allows unmanned aircraft systems (UAS) to fly as long as their operators keep them in sight, fly below 400 feet, and avoid populated areas and airports.

The FAA also forbids any drone to be flown for business purposes. "In the U.S. right now, it's completely open, so long as you do it for noncommercial purposes," Pirker says. "The cool thing is that this is still relatively new. None of the laws are specifically written against or for what we do."

While the FAA did not sanction Team BlackSheep for buzzing landmarks as a publicity stunt, it has shut down other for profit drone operators, including Minneapolis-based Fly Boys Aerial Cinematography, which was using drones to take photographs for real estate developers.

Even the military and other government operators must obtain FAA waivers to operate drones. That means that flying over a wooded area is fine for an amateur, but a fire department that uses a drone to scout a forest fire in the same area requires special federal permission.

Federal, state, and local agencies must apply for FAA waivers to put drones to work. Although the process is cumbersome and time-consuming, there has been a sharp rise in requests. For example, the United States Geological Survey (USGS) operates a fleet of 21 T-Hawks, ducted-fan UAS that can be readied for takeoff in 10 minutes and can ascend to altitudes of 8,000 feet. The USGS has obtained permission to use these craft to view hard-to-reach cliff art, track wildlife, inspect dams, and fight forest fires.

Others are not so lucky. Last year the FAA grounded a $75,000 drone that the state of Hawaii bought to conduct aerial surveillance over Honolulu Harbor. The agency would not waive the rules because the flights were too close to Honolulu International Airport.

In a bid to force a reassessment of the regulations, Congress in 2012 ordered the FAA to open the National Airspace System (NAS) to unmanned aircraft. The law sets a deadline of 2015 for the FAA to create regulations and technical requirements that will integrate drones into the NAS. "Once the rules of the air are established, you're going to see this market really take off," predicts Ben Gielow, a lobbyist for the Association for Unmanned Vehicle Systems International.

The FAA predicts that, because of the law's passage, 30,000 public and private flying robots will be soaring in the national airspace by 2030. For a sense of scale, 350,000 aircraft are currently registered with the FAA, and 50,000 fly over America every day.

Some experts express alarm at the prospect of tens of thousands of extra aircraft flying in the already cramped U.S. airspace. "To most of us air traffic controllers, it's unimaginable. But we're also smart enough to know it's coming," says Chris Stephenson, an operations coordinator with the National Air Traffic Controllers Association. "I refer to UAS as the tsunami that's headed for the front porch."

N MAY 9, 2013, THE ROYAL CANADIAN Mounted Police in Saskatoon, Saskatchewan, were frantically searching for a missing person at the scene of a car accident. They knew the 25-year-old man was out there—the dazed victim had walked away from the scene in just a T-shirt and pants, calling for help on his cellular phone before falling unconscious into the snow.

Stymied, the Mounties called forensic collision technician Cpl. Doug Green and his partner—the Draganflyer X4-ES, a 36-inch-wide quadrotor drone equipped with an infrared camera. Canada, which has been working out detailed regulations for commercial use of UAS since 2008, allows police to operate drones, thus setting the stage for what the Mounties say was the first life-saving rescue by a civilian drone.

Green commanded the small quadrotor aircraft into the air and directed it across the snowy landscape. The forward-looking infrared cameras quickly saw the warmth of the victim's body, 2 miles from the accident scene, and the Mounties moved in to rescue him.

Despite such potential to save lives, government drones can become tools of overreaching law enforcement.

In June the citizens of Iowa City, Iowa, collected thousands of signatures on a petition that successfully pressed the city council to ban license-plate readers and drones. The fact that neither was operating in Iowa City was no matter. The effort is part of a larger trend. This year, at least three bills to restrict the use of UAS in collecting data or conducting surveillance without a warrant have been introduced in Congress, and laws that sharply curb the police use of drones have either passed or have been proposed in 14 state legislatures.

The same things that make drones useful cause the public to worry: UAS are much cheaper to own and operate than helicopters or fixed-wing airplanes. "Drones drive down the cost of surveillance considerably," Ryan Calo, an assistant professor at the University of Washington School of Law, testified before Congress in March 2013. "We worry that the incidence of surveillance will go up."

The International Association of Chiefs of Police last year adopted guidelines stating that law enforcement agencies should obtain search warrants before launching UAS to collect evidence of a specific crime and must delete all unused images.

These steps may not be enough to mollify critics. "Just because the government may comply with the Constitution does not mean they should be able to constantly surveil, like Big Brother," Republican Sen. Charles Grassley of Iowa said at the March congressional hearing.

AVID QUIÑONES TAKES A HOMEMADE QUADCOPTER out of the trunk of his black SUV. The 38-year-old sole proprietor of video production company SkyCamUsa attached a video camera to a swiveling servo and duct-taped it to the quadcopter's fuselage. He carefully lays his drone on the ground of the small parking lot outside the Kissena Velodrome, a cycling track in the New York City borough of Queens.

For six years Quiñones, has been "suffering and waiting" for FAA approval to offer his homemade fleet to advertising, television, and movie productions. His time has nearly come: Next year, the FAA will issue rules allowing electric- powered UAS that weigh 25 to 55 pounds for commercial flights. When that happens, "it'll open up a whole new world of possibilities," Quiñones says.

Better controls, longer ranges, and miniature, high-resolution cameras have revitalized the radio-control-aircraft community. Some of the cornerstone technologies of military drones, such as automatic camera stabilization, the ability to follow waypoints of a flight path, and the automatic process to return home if controls are lost, are now available to RC hobbyists. Many of these amateurs will be trying to turn pro next year.

The FAA rules will accommodate their ambitions, but the agency will also impose restrictions. The FAA is not providing many details, but its regulations will certainly limit when a drone can fly and likely will require licensing of the aircraft and the pilots. Even Team BlackSheep's Pirker argues that there should be a way to distinguish capable RC fliers from amateurs. "There needs to be some kind of process that evaluates if you know what you're doing, something in the line of aircraft or pilot certification," he says.

The impact of the FAA's new regulations is expected to be dramatic. Agency officials estimate that 7,500 small commercial UAS will be operating in U.S. skies by 2018.

HERE'S A HURRICANE FORMING IN the Atlantic Ocean, gathering strength off the coast of Africa. It's time for NASA's Global Hawk to go to work.

Hundreds of miles from the storm, pilots on the ground in Wallops Island, Va., level the 44-foot-long drone at 60,000 feet, aiming it over the top of the hurricane. This altitude is twice as high as the maximum at which manned hurricane chasers operate. From that unique vantage, the Global Hawk aims Doppler radar, radiometers, and microwave sensors into the heart of the storm.

Whereas manned planes usually have to return to base soon after they reach a hurricane developing in the distant Atlantic, the Global Hawk can linger on location for dozens of hours, collecting data that can be used to predict the system's path and severity. Last year, NASA flew one Global Hawk per storm; this year two UAS will study each hurricane from different vantage points.

From the volcanic plumes of Costa Rica to the icy north Atlantic, drones go where sensible pilots cannot or will not go. The larger and more powerful the aircraft, the more payload it can carry and the farther it can travel. And drones with gas-powered engines can stay aloft longer than their smaller, battery- powered brethren or similar size manned aircraft.

Oil company engineers planning the least environmentally risky route of a

A Gallery of Drones

Volcano Diver

NAME/MAKER: Dragon

DESCRIPTION: Drones can collect previously inaccessible data for scientists. This year NASA launched 10 drone flights into the plume of Costa Rica's Turrialba volcano to measure ash and gas concentrations. The data will help build simulations that will safeguard global airspace during future eruptions.

Crowd Controller

NAME/MAKER: Hover-Mast/Sky Sapience

DESCRIPTION: This tethered flier quietly rises 165 feet from a truck or boat to monitor the area below with high-definition and infrared cameras, laser designators, and gear to snoop on electronic communications. Its stealthy operation and absence of registration as an aircraft (because it's tethered) raises civil libertarians' hackles.

Smoke Jumper

NAME/MAKER: Flanker/FireFlight Unmanned Aircraft Systems

DESCRIPTION: Tough jobs require tough tools. The hand-launched Flanker, made with durable expanded polyolefin foam and carbon-fiber spars in the wings and canopy, can survey a fire with infrared cameras. The shrinking size of imagers transforms small drones into great fire-fighting tools.

Terrorist's Toy

NAME/MAKER: Model F-86 Sabre/E-Flite

DESCRIPTION: In 2011, Rezwan Ferdaus attempted to stuff C-4 explosives into 3-foot-long RC planes and fly them into Washington, D.C., buildings. Snared by an undercover FBI agent, Ferdaus is spending 17 years in prison. The difference between a toy, a drone, and a guided missile is largely determined by the user's intent.

Rescue Robot

NAME/MAKER: Draganflyer X4-ES/ Draganfly Innovations

DESCRIPTION: The Royal Canadian Mounted Police used this 36-inch wide quadrotor's forward-looking infrared camera to find a crash victim. The Mounties claim this is the first life saved by a civilian drone. First responders appreciate having access to—and control over—aircraft, rather than relying on a limited number of helicopters.

→ Regulating America's Air

THE SKY OVER AMERICA IS CARVED INTO FEDERALLY REGULATED ZONES, CALLED CLASSES, EACH WITH ITS OWN SINGLE LETTER DESIGNATION AND RULES.

IF YOU INTEND TO BUILD AND FLY A DRONE RECREATIONALLY, YOU FACE ONLY A FEW RESTRICTIONS, WHICH COME TO YOU COURTESY OF THE FEDERAL AVIATION ADMINISTRATION. There is currently no legal way to operate drones for profit; however, changes are on the way. Congress has directed the FAA to devise rules by late 2015 in order to integrate UAVs into the nation's airspace, with an earlier deadline of August 2014 to formulate regulations for small, recreational UAVs weighing less than 55 pounds.

Airspace: CLASS G
Altitude: 0-1200 feet

The FAA anticipates that within five years 7,500 small commercial drones will be operating at these low altitudes. Current rules allow radio-control hobbyists to fly their aircraft within sight and under 400 feet. These rules, detailed in FAA Advisory Circular 91-57 and published in 1981, were written for model aircraft, but for now the FAA is applying the same rules to UAVs (see "Drone Skies" on page 64). However, a new generation of small drones can go higher and follow GPS waypoints beyond a controller's visual range, raising worries about midair collisions with helicopters. Expected FAA restrictions may limit commercial drone flights to daytime hours, keep them away from helipads, and require an operator's license.

Airspace: Airports
(Class B, C, D)
Altitude: 0-10,000 feet

Airspace surrounding towered airports consists of one or more tiered layers. The highest layer of the biggest (Class B) airports has a maximum radius of 30 nautical miles. FAA regulations require all aircraft operating in these airspaces to be equipped with two-way communication for exchanges with air traffic control (ATC). Even drones with a high level of autonomy will likely need to take off and land under ATC supervision.

new pipeline, for example, would not be interested in a radio-controlled T-Hawk. They'd want a fixed-wing aircraft the size of a Cessna to efficiently scan the terrain below in a single flight using 3D-radar-mapping sensors.

Congress is demanding that the FAA develop a plan to govern larger unmanned aircraft by September 2015. In theory, everywhere we today see a helicopter or private airplane, there could be a drone. Future operators of FAA-certified unmanned aircraft could simply file a flight plan before takeoff, as pilots of manned aircraft do. But the FAA's rules will likely include prohibitive technical regulations.

Coming up with safety standards for large drones that fly at thousands of feet is far more complicated than regulating the operation of an RC plane zipping along within the operator's line of sight. The current standards for manned aircraft require pilots to "see and avoid other aircraft." Without humans on board, flying robots will need to think for themselves to avoid other airplanes.

Aerospace companies and the Pentagon are developing systems that combine radar, cameras, or other sensors with software that will detect aircraft and change course to avoid them. Some of the systems rely on ground stations, while more advanced versions are incorporated into the drones.

This solution comes with engineering drawbacks, however. "By hanging that type of technology on an unmanned aircraft, you start adding a lot of weight and draining a lot of power," says Viva Austin, the civilian official in charge of the Army's ground-based sense and avoid project.

Airspace: Class A
Altitude: 18,000–60,000 feet

This highly regulated slice of sky incorporates jet routes, where jet-powered drones could share the sky with airliners, military transports, and general aviation craft. As with manned aircraft, drones will have to carry transponders that provide location and altitude to air-traffic radar. In Class A airspace, pilots mainly rely on instrument readings rather than visual cues, except in the case of collision avoidance. Because of this, unmanned aircraft will need yet-to-be-developed sensors that enable them to detect and avoid other airplanes autonomously during an emergency.

Airspace: Class E
Altitude: All undefined airspace below 18,000 feet; everything above 60,000 feet

The upper fringes of airspace hold a lot of promise for pilotless craft. Here, a drone could fly along an ATC-approved flight path without further communication, as long as it had a transponder and stuck to its course. High-flying drones are designed primarily for endurance, not speed, so the wake turbulence caused by faster manned airplanes could endanger them. Military jets also operate here, so high-flying drones would have to be able to sense and avoid supersonic aircraft—a big hurdle, considering how quickly a drone would have to react.

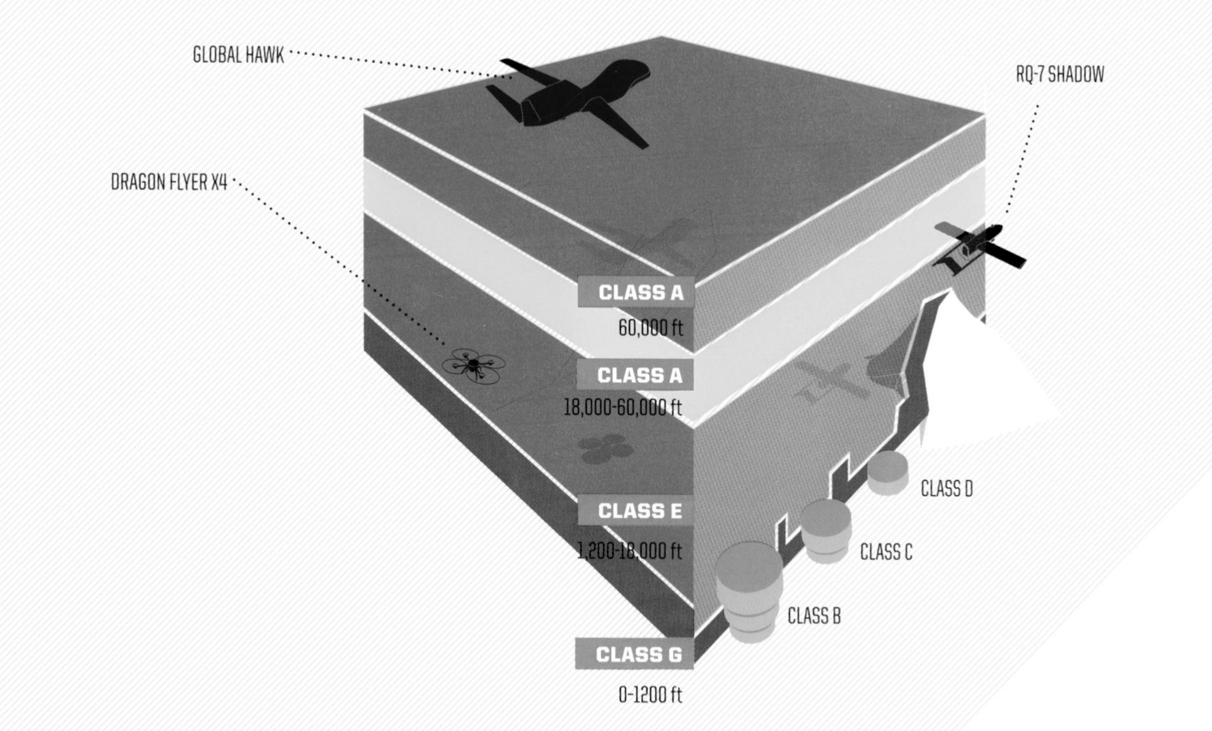

John Walker, a former FAA director and cochair of a federal advisory panel that is developing standards for UAS technology, says technical demands will likely slow the pace of drone adoption. For example, the panel may recommend that the FAA require sense and avoid systems that will steer a drone away from potential collision courses, not just perform the simple "climb or descend" instructions current systems give a pilot.

That requires a flight-control computer powerful enough to handle complex algorithms. "What we're talking about for separation assurance is climb, descend, turn left, turn right," Walker says. "It's going to take a tremendous amount of modeling and simulation."

The result? Walker predicts manufacturers and operators will have to invest a lot of money and years of work to meet the pending FAA requirements. Once the tech is developed, the feds will test and certify it, causing more delays.

EDITOR'S NOTE

In June of 2014, the *Washington Post* reported that a government audit revealed that the FAA is significantly behind schedule to determine rules for integrating drones with standard U.S. air traffic. In the absence of these rules, a de facto ban on commercial drones will persist. —DHW

→ Big Future for Little Fliers

DON'T LET THE SIZE OF MICRO AIR VEHICLES, (MAVS) FOOL YOU. Although some measure less than a foot across, these remote-control reconnaissance planes have proven to be deft, powerful tools for soldiers and police operating in the narrow airspace of cities and towns. But diminutive aircraft can't carry much fuel or many batteries, so their use is often limited by short flight times. With civilian and military officials clamoring for more advanced urban MAVs, defense firms are developing models that can undertake longer surveillance missions. Future MAVs will also be able to set down on rooftops or ledges, which consumes less power than hovering and maintains a more stable viewing platform for cameras.

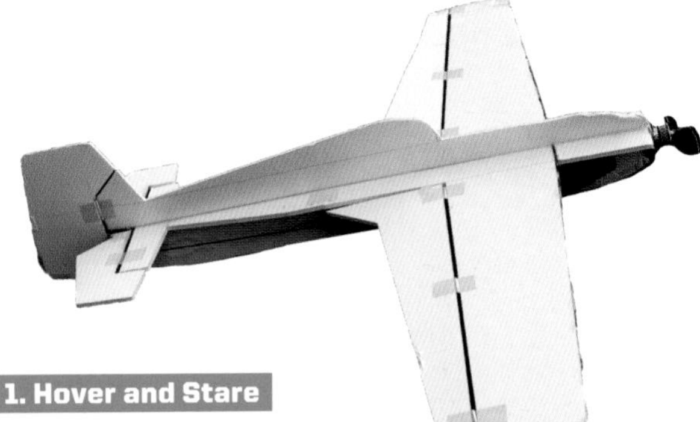

1. Hover and Stare

CYBERQUAD
LENGTH: 21"
MAX WEIGHT: 6.6 LB

The electric CyberQuad, built by the Australian company Cyber Technology, uses four ducted fans that allow it to fly and hover. Like a helicopter, it can set down on any flat surface. Shrouds around the rotors give operators confidence to fly the craft in areas where an exposed blade might catch—a feature that came in handy during the recent inspection of a damaged oil rig in the Timor Sea.

2. Stand Up and Fly

XQ-138
LENGTH: 28"
MAX WEIGHT: 6.2 LB

The XQ-138, designed by University of Kansas aerospace engineering associate professor Ron Barrett, takes off and lands on its tail, courtesy of a spinning rotor. The entire airframe of the XQ-138 rotates 90 degrees for 150-mph forward flight. Actuators counter gusts of wind to keep the craft stable.

What the coming drone invasion will look like is still uncertain. Even the FAA is unwilling to predict how the regulations will shape the future of flight. In a written response to *Popular Mechanics'* questions, the FAA said that "until the plan is complete, the exact extent of UAS operations proposed for 2015 is difficult to predict."

Despite the assured language of the law, the integration of UAS into America's airspace can't be done with the stroke of a pen. That might be seen as a good thing by those who are nervous about the coming wave of unmanned aircraft. But one day the technology and the laws will be ready, and the skies will be fully open to robots. How they are used will be up to us.

EDITOR'S NOTE

As of publication, drone use is still outstripping the FAA's ability to regulate. Drones recently employed by wedding photographers, real estate brokers, the San Jose Police Department, as well as drug smugglers moving methamphetamine across the border between the United States and Mexico. —DHW

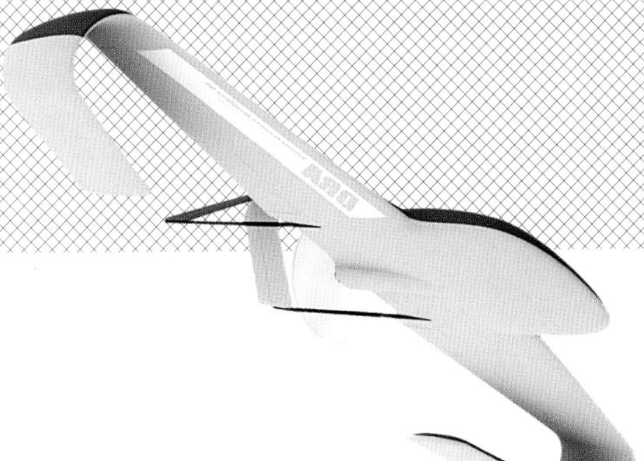

3. Stick Around

STANFORD DRONE
LENGTH: 39"
MAX WEIGHT: 0.8 LB

At Stanford University's Biomimetics and Dexterous Manipulation Lab, professor Mark Cutkosky is developing a drone that can land on walls using sticky feet. The plane executes a deliberate stall as it approaches a wall and turns its underbelly to face the vertical surface. Flexible legs absorb the shock of impact. Tiny claws on the feet give the aircraft a tenacious grip, allowing it to crawl around and reorient before taking off again.

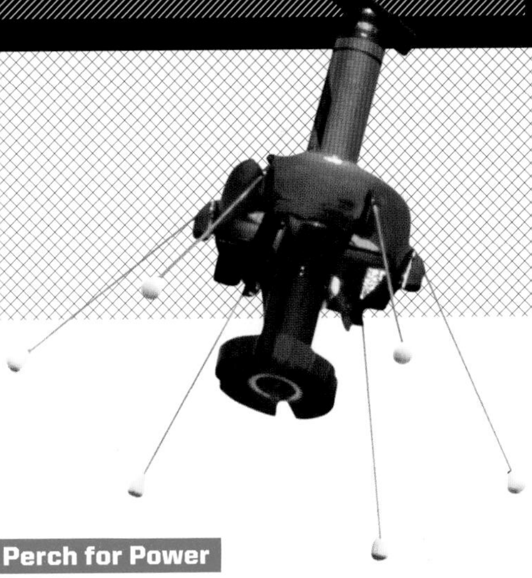

4. Perch for Power

DEVIL RAY
LENGTH: 24"
MAX WEIGHT: 2.5 LB

Engineers at Defense Research Associates designed a specialized perching craft for the Air Force. The Devil Ray can cruise at 42 mph, but its upside-down, turned-in winglets give it stability at slow speeds. The MAV's precise control is needed to snag power lines with a custom-made latch. Perched like a raven, the Devil Ray then tops off its energy with an induction-based recharger.

EVEN *AS THE AIR FORCE FRANTICALLY EXPANDS ITS FLEET OF MQ-9 REAPERS, THE SERVICE IS SEEKING A TOUGHER, FASTER, AND SMARTER SUCCESSOR. "WE ARE GOING TO REPLACE THEM BEFORE THEY FAIL," SAYS THE WING COMMANDER IN CHARGE OF THE REAPERS.*

The U.S. military must evolve to fight, and that means choosing which new Unmanned Aerial Systems (UAS) will be incorporated into its armed forces.

BY **JOE PAPPALARDO**

This article was published in *Popular Mechanics* in 2010.

IKE ITS WATERFOWL NAMESAKE, THE HERON unmanned aerial vehicle has the excellent vision of a hunter. Today, the 27-foot-long Israeli UAV is making a rare flight over the United States, using a high-definition video camera to track a speedboat buzzing across the Patuxent River in Maryland. The camera shares space with an infrared thermal imager and laser rangefinder inside a 17-inch sphere mounted under the aircraft's nose. The camera and the UAV both turn automatically to track the boat below, no satellite-linked joysticks required. On the Patuxent, a Coast Guard crew in a shallow-water patrol boat uses a real time video feed from the Heron to locate the speedboat.

Less than 5 miles away, several hundred spectators watch the camera's feed on a massive color television monitor. The crowd of defense officials, defense industry wonks, and military aviation buffs–many with bumper stickers on their cars that say "My other vehicle is unmanned"– is thick here at Webster Field, an auxiliary naval airfield in Maryland. The Heron is just one of about a dozen UAVs making flight demonstrations. As each one sweeps overhead, an announcer gushes over its abilities with the over-enthusiasm of a county fair emcee describing a prize sheep.

The crowd watches on the massive screen as the two boats converge and the Coast Guard crew completes the mock interception. The image of the river scene wheels as the Heron banks away from the boats and returns to the airfield. The UAV glides into a smooth, autonomous landing and as the Heron taxis, the goofball emcee coos over the PA speakers, "Aw, isn't that just pretty?"

The day is a spectacle of flying robots. A unit of Textron shows off an aircraft that it is pitching to the Marine Corps. It has a 12-foot wingspan and a pusher propeller mounted between its fuselage and inverted V-tail. It can be launched from a moving vehicle and is recovered by flying it into a net. The U.S. Army also has a marquee UAV to demo, the MQ-8B Fire Scout. The 3,150-pound unmanned helicopter, the Army's first, may soon scan battlefields for chemical weapons, minefields, and radio transmissions. And the showstopper, even while remaining earthbound, is the Navy's Joint Unmanned Combat Air System, a sleek, blended-wing aircraft with the maw of an air inlet placed almost mockingly where a cockpit would go. It sits like a resting bird, its 31-foot-long wings folded up for better storage on a warship. It is scheduled to perform an autonomous takeoff and landing from an aircraft carrier deck this year.

"TODAY, THE EVOLUTION OF THE MACHINE IS BEGINNING TO OUTPACE THE CAPABILITY OF THE PEOPLE WE PUT IN THEM... WE NOW MUST RECONSIDER THE RELATIONSHIP."

—U.S. AIR FORCE CHIEF OF STAFF GEN. NORTON **SCHWARTZ**

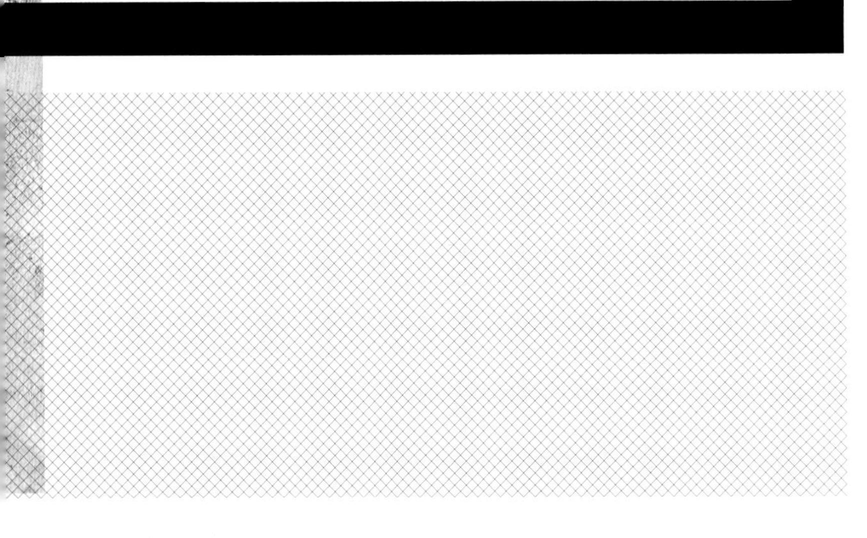

With all the hardware and enthusiastic attendees, it's easy to overlook a missing guest—the U.S. Air Force. Of all the advanced aircraft on the flight line, none is being developed for Air Force programs or controlled by the service's airmen.

Unmanned aircraft are the biggest thing to happen in military aviation since stealth geometry, and the Air Force's leadership is dramatically increasing the UAV fleet this year. However, the service is still struggling over how the technology can be maximized in the future. "Today, the evolution of the machine is beginning to outpace the capability of the people we put in them," Air Force chief of staff Gen. Norton Schwartz said late last year in a speech to the Air Force Association. "We now must reconsider the relationship."

Under his direction, the Air Force is trying to become the Pentagon's leader of future UAV development. Schwartz's primary tool is the "Unmanned Aircraft Systems Flight Plan, 2009-2047," a comprehensive look at how the U.S. military can expand the use of UAVs over the next 38 years. The Air Force is proposing to use next-generation unmanned aircraft in a slate of new missions, including air strikes, aerial refueling, cargo transport, and long-range bombing.

But how much freedom will the Air Force be willing to grant unmanned airplanes? Its airmen are only now coming to accept UAVs—they fly them every day over Iraq, Afghanistan, the Horn of Africa, and other hot spots. The service has articulated a way forward that not only marginalizes pilots, but also promises to replace many UAV ground control crews with automation. Today's highly trained airmen may not embrace this vision of the future. One Air Force officer working with unmanned aircraft would only say he supports the report "because it's a plan. And having a plan is better than not having a plan."

THE AIR FORCE SQUANDERED DECADES' worth of opportunities to lead U.S. military UAV development. In the 1970s, the service experimented with unmanned surveillance craft in Vietnam but dropped all funding after it decided the technology did not offer improvements over traditional airplanes. Continued advances of Soviet warplanes, such as the MiG fighter, kept a Cold War premium on air superiority won by high performance, expertly piloted airplanes.

The idea of unmanned airplanes also runs contrary to the airman-centric ethos that has defined the Air Force since it became an independent military branch in 1947. *Aviation Week and Space Technology* magazine in 1973 quoted an Air Force official's disparaging verdict on remote-control warplanes: "How can you be a tiger sitting behind a console?" That attitude proved to be shortsighted. In 1982, Israel used UAVs to spoof Syrian radar in Lebanon, but the status quo in America continued for another decade. The Pentagon started UAV research in the mid-1990s, but even then, the funding was tepid—in part because of Washington's bias toward large, job-generating manned airplane programs.

Guerrilla wars in Iraq and Afghanistan changed all that—the need for constant overhead video is driving a UAV spending spree. When facing insurgents who blend into a local population, good intelligence is worth more than even the smartest bomb. In 2010, the Defense Department will spend $5.4 billion on unmanned aircraft development, procurement, and operations—about $2.5 billion more than the military spent on UAVs during the 1990s.

Experts Weigh In

Guy Ben-Ari

Senior policy analyst, Center for Strategic and International Studies

"I think the Flight Plan is a serious document. It's not just discussing the technology, but the policy, the legislation, the ethical framework. The whole package needs to be developed in parallel as these technologies mature."

P.W. Singer

Author, *Wired for War*, The Brookings Institution

"The road map to 2047 will likely be good for just a few years. But that's all we need for it to make a big difference."

Jim Dunnigan

Author, analyst, strategypage.com

"The other services are pushing ahead with their UAV efforts without paying much attention to the Air Force. No one has any idea what the tech will be in 2017, much less 2047. In 2047, we'll have stuff as unfamiliar to us as today's tech would be to someone in the late 1940s."

This boom is causing turf wars within the Pentagon. Military branches seldom develop weapons systems together, despite the potential savings of time and money if the services shared research costs and ordered hardware in bulk. The Air Force wants to coordinate UAV development within the Pentagon and drafted its ambitious Flight Plan to describe how the service would serve as the Pentagon's chief guide to unmanned airplane development, in concert with the Army, Navy, and Marine Corps. "The Flight Plan is part of an Air Force effort to lay claim over everything that flies, whether it has a pilot or not," says military analyst and author Jim Dunnigan.

The Unmanned Aircraft Systems Task Force, which drafted the plan, is headquartered in a modest office that takes up a small fraction of one floor inside a banal building in Crystal City, Virginia. The full-time staff here tops out at a handful, but National Guard and Air Force Reserve temps fill out the administrative positions. Dozens of moonlighting planners from the Pentagon also volunteer for the task force, forgoing their free time for a chance to work on a project with high-ranking luminaries at Air Force headquarters who advise the task force.

The day-to-day work is supervised by the task force's director, Col. Eric Mathewson. The former F-15 pilot is a compact man with a soft, smooth voice that always sounds earnest. Mathewson often places a hand on his head when he speaks, as if his ideas could burst from his temple, if he weren't holding them in. "It was clear we had been reactive, reactive, reactive," Mathewson says. "It was time to develop a vision."

That vision depends on developing smarter unmanned aircraft that can make life-and-death combat decisions on their own. According to the Flight Plan, UAVs will demonstrate "sense and avoid" collision-avoidance systems by the end of this year. Unmanned aircraft will be able to refuel each other by 2030. Global strike capability, perhaps even with nuclear weapons, is projected for 2047. "As technology advances, machines will automatically perform some repairs in flight," the Flight Plan reads. "Routine ground maintenance will be conducted by machines without human touch labor." The Air Force document not only discusses once-taboo subjects, such as automatic target engagement and autonomous UAVs flying in commercial airspace, it also includes short-term recommendations and goals to one day make them feasible.

Mathewson says that by 2020 just one control crew—airborne or groundbased—will be able to control multiple UAVs at once. Ground-control crews today, even when aided by advanced autopiloting, continuously monitor a single UAV. This level of direct control and supervision is referred to as man-in the- loop. But a robotic system that only alerts humans when a critical decision needs to be made is called man-on-the-loop. A ground-control crew can opt to redirect the UAV or assume direct control until the key choice is made. "I don't think it's an overstatement that this is a revolution of military affairs," Mathewson says. "The revolution is the conscious application of automated technology."

"I DON'T THINK IT'S AN OVERSTATEMENT THAT THIS IS A REVOLUTION OF MILITARY AFFAIRS. THE REVOLUTION IS THE CONSCIOUS APPLICATION OF AUTOMATED TECHNOLOGY."

—COL. ERIC MATHEWSON, UNMANNED AIRCRAFT SYSTEMS TASK FORCE DIRECTOR

AN-ON-THE-LOOP CONTROLS could make a battlefield look like this: An F-35A Lightning II fighter cuts through the night sky. The pilot's mission is simple—destroy an enemy bunker protected by a network of radar and antiaircraft missile batteries. His three wingmen—one flying scant feet away, another 150 miles ahead and the third preparing to cause a diversion far to the east—are following a meticulous battle plan meant to defeat these defenses. Of the four aircraft in the strike group, only the F-35A has a cockpit; the rest are semiautonomous UAVs that the pilot must trust with his life.

→ Tiny Drone Squadron

SWARM MICRO-UAVS: VIJAY KUMAR, SHAOJIE SHEN, MATTHEW TURPIN, DANIEL MELLINGER, AND ALEX KUSHLEYEV (UNIVERSITY OF PENNSYLVANIA); NATHAN MICHAEL (CARNEGIE MELLON UNIVERSITY).

VIJAY ***KUMAR, DANIEL MELLINGER, MATTHEW TURPIN, AND ALEX KUSHLEYEV (FROM LEFT), PHOTOGRAPHED BY NATHANIEL WOOD AUGUST 16, 2013, AT PENN'S GRASP LAB.***

One of the most dangerous missions in military aviation is suppression of enemy air defenses (SEAD). The lead UAV becomes bait as it flies into radar range of antiaircraft missile batteries. An icon on the F-35 pilot's virtual headup display, projected onto the faceplate of his helmet, alerts him that the SEAD unmanned airplane has automatically identified the emissions of an enemy radar site. This is the first time in the mission that the SEAD airplane has communicated with any human.

Miles from the danger, the F-35A pilot coolly assesses the situation displayed on one of the screens in his cockpit, confirms the target is legitimate, and authorizes the lead UAV to fire. The AGM-88 high-speed antiradiation missile follows the radar waves back to their source, obliterating the dish and its crew. There is now a gap in the enemy radar screen, and the pilot directs the UAV to return to base.

Meanwhile, another UAV east of the target, navigating by using a mix of GPS and accelerometer data, is busy scrambling other enemy radar installations by flooding the skies with emissions that share the radar's frequency. The jamming pods under the UAV's wings also disrupt radio transmissions from the air-defense network, covering up the sudden loss of contact with the radar sites protecting the bunker. Otherwise, an enemy commander could discover the location of the actual raid. After a preset amount of time spreading confusion, the UAV returns to base.

The F-35A pilot is closing in on the target fast and needs to carefully aim the F-35's electro-optical targeting system to release a bomb that will hit the structure at an angle calculated to collapse it without destroying nearby civilian buildings. He triggers the laser designator and authorizes the nearby unmanned airplane to drop a pair of bombs, which use fins to steer toward the laser-designated sweet spot. The pilot watches the twin, concurrent explosions, makes a quick battle-damage assessment, and, satisfied, banks the airplane and heads back to base. His robotic wingman follows his lead, flying evenly at his side.

T CAN BE HARD TO SEE THE FLIGHT PLAN'S vision of autonomous flying robots from the human-intensive work being done at Creech Air Force Base in Nevada. The desert base is in the midst of an unprecedented boom as it hosts the fast-growing 432nd Air Expeditionary Wing, the only one dedicated solely to flying unmanned aircraft.

Every aircraft and satellite-linked ground-control station here is being used to fly missions in the Middle East, the Horn of Africa, and points beyond. New buildings fill up with staff as soon as the construction dust settles. "Every time the fishbowl grows, the fish get too big for it," says Col. Pete Gersten, the 432nd's commander. Mathewson served at Creech as group commander before Gersten's arrival, but their jobs now are pointed in opposite directions. As Gersten wrestles with recruiting ground-control crews, Mathewson promotes ways to replace the airmen with artificial intelligence.

Every time an airman is replaced by a machine, the Air Force cuts the cost of health benefits, base upkeep, and recruitment. Current unmanned systems require as many, if not more, people to fly missions than piloted airplanes do. For example, it takes a crew of three to operate a Reaper, even while it's on autopilot: One to fly, another to operate the sensor ball in its nose, and a third to serve as military intelligence liaison. Another pair must deploy to the forward airfield to guide the UAV, using line-of sight radio during takeoff and landing. By replacing these positions with automated functions, the cost of joystick operators could plummet.

But Gersten–who calls his unmanned airplanes "remotely-piloted vehicles" to emphasize the crews operating them–does not give up human control over the aircraft unless it provides a clear war-fighting edge. For example, the Flight Plan pegs autonomous takeoff and landing for the Reaper by the end of 2010, but Gersten is not begging for that ability. In fact, when faced with a rash of accidents during landings, Gersten chose a solution to help, not replace, the joystick pilot.

CREECH *AIR FORCE BASE, NEVADA, HAS THE ONLY WING DEDICATED TO UNMANNED AIRPLANES LIKE THE MQ-9 REAPER (SHOWN). GERSTEN IS EAGERLY SEEKING CREWS TO OPERATE UAVS, BUT ISN'T READY TO REPLACE THEM WITH SOFTWARE.*

The Replacements

THE AIR FORCE ENVISIONS SWAPPING ITS PILOTS FOR A FLEET OF VERSATILE AND AFFORDABLE UNMANNED AIRPLANES. A SINGLE UAV WITH INTERCHANGEABLE PAYLOADS COULD REPLACE SEVERAL LEGACY AIRPLANES. HERE'S A LOOK AT SOME POSSIBLE TRADES.

PRESENT	FUTURE
F-16 FIGHTING FALCON This oft-upgraded multirole warplane has proved itself in dogfights and air strikes since 1979. **MC-12W LIBERTY** In 2009 this plane began flying battlefield surveillance missions.	**MULTIMISSION UAV** Medium-size UAVs will swap onboard gear and weapons to intercept communications, bomb ground targets, or fight enemy aircraft. This year, the Pentagon will select a design for a 2015 replacement of the MQ-9 Reaper.
KC-135 STRATOTANKER This 136-foot airplane can offload 6,500 pounds of jet fuel per minute but fills only one airplane tank at a time. The average age of the Air Force's fleet of tankers, flying since 1957, is now more than 40 years.	**JOINED-WING AERIAL REFUELING UAV** A box-wing UAV could fuel many airplanes at the same time and loiter, perhaps for a week, until needed. The Pentagon is spending more than $40 billion on manned refuelers, but unmanned tankers could be built to service UAVs.
B-2 SPIRIT Since 1989, this stealth bomber's mission has been to attack well-guarded ground targets. **U-2 DRAGON LADY** This unarmed, high-altitude recon airplane, in service since 1957, can fly 12-hour missions.	**LONG-RANGE SURVEILLANCE BOMBER** This stealth UAV could monitor a target for days—and then destroy it at the time of a commander's choosing. The Air Force hopes to restart its bomber program this year. The new aircraft will likely be able to fly with or without a pilot.

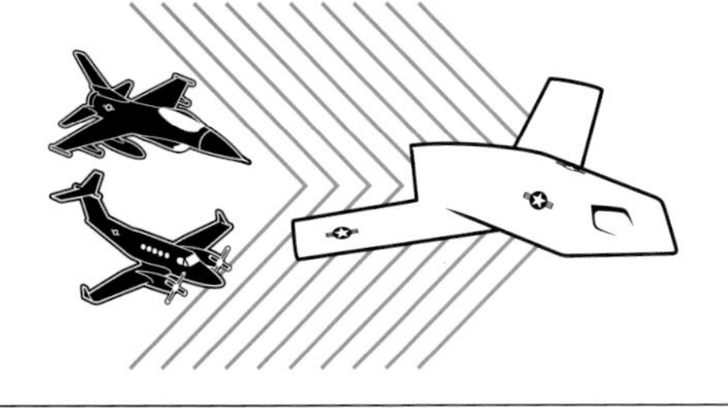

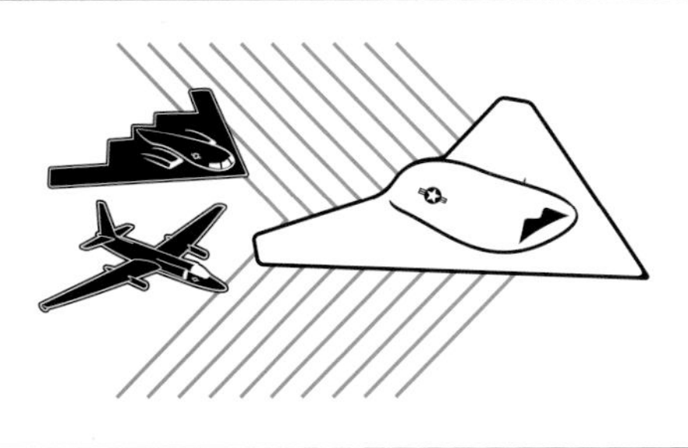

WHEN *UNMANNED AIRCRAFT CAN REFUEL ONE ANOTHER, THEIR TIME ON A MISSION WILL BE DRAMATICALLY EXTENDED. THE AIR FORCE RESEARCH LABORATORY WILL SPEND $49 MILLION OVER THE NEXT FOUR YEARS TO CREATE A SYSTEM ALLOWING UAVS TO AUTONOMOUSLY REFUEL IN THE AIR, AS SEEN IN THIS 2007 PREDATOR TEST.*

→ Future of Air Defense: Laser-Drone Wars

FOUR YEARS AGO, THE AIRBORNE LASER TEST BED, A MODIFIED 747 THAT SHOOTS LETHAL PHOTONS, DESTROYED A BALLISTIC MISSILE IN FLIGHT BY BLASTING IT WITH A CHEMICAL LASER. But after its test demonstration, the Pentagon mothballed the plane because of its $100-million-a-year price tag. Still, the dream of flying laser weapons won't rest in the boneyard, if DARPA has anything to do with it. The agency is researching fiber-optic lasers as an alternative to massive chemical lasers. Lockheed Martin and Northrop Grumman were awarded contracts to build a prototype for a classified program called Endurance. Instead of blowing up a missile, this system will target optically guided missiles. The program can take advantage of lightweight lasers on drones, because it takes less power to scorch optics than to melt a rocket casing. This technology could protect combat aircraft or counter terrorist threats to commercial airlines.

A. COMBATANTS FIRE SURFACE-TO-AIR MISSILE AT A 747.
B. THE DRONE FIRES A LASER BEAM AT THE MISSILE.
C. THE LASER STRIKES THE MISSILE'S OPTICS.
D. THE MISSILE IS DEFLECTED.

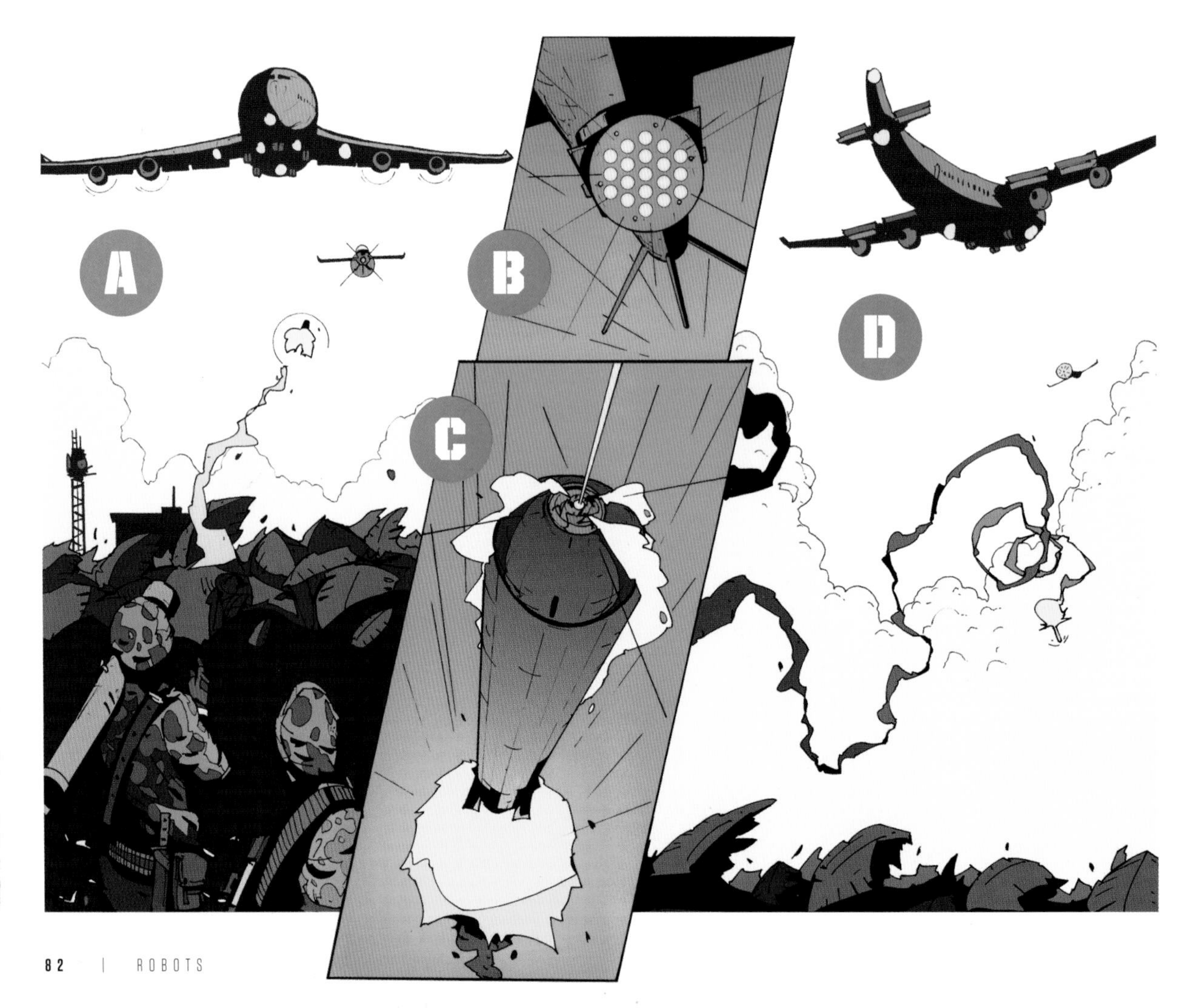

The landing gear would collapse when Gersten's UAVs bounced down the runway, making it difficult to find the correct pitch of the nose and causing oscillations that can destroy the aircraft. The seemingly obvious solution: Program the machines to take over and land automatically—something the Army's Sky Warrior, which is nearly identical to a Predator, already does. But Gersten opted for a simpler fix, adding a triangular carrot icon on the flight-control screen that sets the correct pitch to prevent the oscillation cycle from starting. This change will be made to ground-control stations this year, and he says "the cost is minuscule."

Gersten's reaction to the Flight Plan is coolly receptive. (He rolls his eyes at the report's language that suggests that UAVs could one day carry nuclear weapons.) The lower ranks on the base are more frankly skeptical of autonomy. Senior Airman Jessie Grace, a sensoroperator instructor at Creech, has spent wrist-aching hours keeping a UAV's camera trained on a target vehicle or locking his tired eyes on display screens to catch subtle signs of insurgent activity. While he does say that pilots could control more than one airplane at once, Grace sees things differently when it comes to his specialty. "I can't imagine a computer doing intelligence, surveillance, and reconnaissance better than a person," he says.

"UNMANNED AIRCRAFT SYSTEMS (UAS) WILL FLY AUTONOMOUSLY TO AN AREA OF INTEREST WHILE AVOIDING COLLISIONS WITH OTHER UAS IN THE SWARM. THESE UAS WILL AUTOMATICALLY PROCESS IMAGERY REQUESTS AND WILL 'DETECT' THREATS AND TARGETS THROUGH THE USE OF ARTIFICIAL INTELLIGENCE."

—U.S. AIR FORCE UAS FLIGHT PLAN, 2009–2047

Mathewson lists battlefield demands as the biggest hindrance to the Flight Plan, but he notes inflexible attitudes as another roadblock. "You see a cultural resistance," Mathewson says. "It's the same thing with the horse cavalry during the introduction of the tank."

NTIL THE FLIGHT PLAN, IT WAS NEARLY impossible to find officials who would even discuss the possibility of unmanned airplanes firing their weapons without human permission. But the report states that by 2030, flying robots could be programmed with "automatic target engagement" abilities. A UAV would open fire only after clearing a checklist of technical details from its sensors—its preset rules of engagement. Such a system would be an heir to ones currently used in Patriot antiaircraft batteries and some antimissile weapons on Navy ships. The legacy of the Patriot is mixed. During the second Gulf War, the system downed a pair of friendly airplanes, killing one American and two British pilots, after mistaking the planes for enemy missiles. Many military officials faulted an over-reliance on automation, but think-tank analysts noted that a lack of training caused the dependence and was the root cause of the tragedies.

Mathewson says that keeping people directly involved at the end of the kill chain is optional but preferred. "There are not that many cases where you'll have free fire, where you're going to have the system completely automated," he says. "If you look at the way we employ unmanned aircraft in the current fights, the rules of engagement require that someone [in charge at the rear] has to approve it, to say, 'Yes, indeed, you're cleared hot' for every single case. And that would hold true."

While Gersten normally keeps any pride in check, the former F-16 pilot can be moralistic in arguing to have a man at the helm of a system that can bring death to its targets. "Warfare should be humanistic," he says. "Human value requires a human interface." It's his way of saying that even sworn enemies deserve to have an actual person, rather than an algorithm, make the decision to kill.

Chapter 04

ROBOTS IN THE... MILITARY

THE NEW ARSENAL

➜ In a world of shifting alliances and constantly evolving strategies, the military has developed a new set of tools for the future.

BY **JOE PAPPALARDO**

This article was published in *Popular Mechanics* in 2014.

OBOTS.

They're the future of war—that's what the Pentagon believes, anyway. Military scientists are creating automated systems that are meant to augment or even replace American soldiers, sailors, Marines, and airmen. China fielding dangerously quiet subs? Make an unmanned subhunting ship to track them relentlessly. Russia selling antiaircraft radar and missile sets to hostile nations? Deploy aerial drones to jam them with electronic attacks. Roadside bombs severing supply lines? Send flocks of unmanned helicopters to deliver supplies.

The robot revolution is also regarded as the way to handle looming budget cuts. While automated systems are expensive to develop, they're cheaper to support than humans—they don't need training, healthcare, or condolence letters.

Not every military robot is a mobile war machine. Some are more subtle, like the automatic-aiming rifle scope on page 91. In the hands of a sniper, it reduces the size of a shooting team by half. But to replace humans entirely, robots must be freed from human control. "Future unmanned systems must be more autonomous, placing less demand on communications infrastructure and shortening decision-making cycles," Air Force Chief of Staff Gen. Mark Welsh III says. The true value of a robot is not that it can be flown remotely, but that it can make the kinds of snap decision pilots face.

Robots can't quite keep up with ground troops, either. Once they can, the Army will be able to "decentralize its formations, making actions less predictable and less vulnerable to asymmetric efforts," says Lt. John Burpo, an advisor in West Point's department of chemistry and life sciences. In Afghanistan, robot helicopters have already proved their worth on resupply missions, protecting Marines from roadside bombs.

Future battlefields will be challenging for man and machine. By 2030, three out of five people in the world will live in urban areas. Entire wars may be fought in cities. "Precision munitions, nonlethal munitions, and physical blockades could play a large role in complex environments," says Lt. Col. Bruce Floersheim, West Point civil and mechanical engineering professor. Weapons like the smart scope could reduce collateral damage—its Heads Up Display boosts first-shot success at targets up to 12 football fields away, and snipers can use smart devices to fire remotely.

At sea, the U.S. is working to keep shipping lanes open. The rise of the Chinese military in the Pacific has brought navy-on-navy battle tactics to the fore. The Chinese are employing a strategy the Pentagon calls Anti-Access/Area Denial—a barrage of land- and air-fired missiles that would keep American

Wearable Robotics: Unleashing Power Suits

RESEARCHERS AT RAYTHEON CONTINUE TO INCH TOWARD THE ARMY'S GOAL OF CREATING ROBOTIC EXOSKELETONS THAT COULD ENHANCE THE STRENGTH AND ENDURANCE OF SOLDIERS. The XOS 2, like an earlier prototype, allows its wearer to lift 200 pounds several hundred times without tiring and to "punch through 3 inches of wood." The newest version of the hydraulically powered suit is lighter, faster, and uses 50 percent less power than the earlier version. That's a key metric, because the researchers currently have to plug in the suit to operate it; a nontethered suit will require further power reductions. Raytheon expects the tethered model to see action in a logistical role, hoisting heavy supplies, in about five years. A nontethered version could follow by 2020.

ships away from coastlines. In response, the Navy is developing ship-launched cruise missiles that can autonomously identify and destroy targets up to 500 miles away.

And then there are the Navy's drones, which recently proved they could handle the toughest job in aviation: taking off from and landing on the pitching, rolling flight deck of an aircraft carrier. The Unmanned Carrier-Launched Airborne Surveillance and Strike (UCLASS) robotic aircraft will increase a carrier group's striking power from 500 nautical miles to 1,500 nautical miles, putting the ships out of range of many land-based weapons. The Navy will choose a UCLASS design from competing vendors as early as next year.

The Navy is also responding to low-tech threats. Diesel-electric submarines can lurk undetected in shallow water, where sonar is less effective, and menace carrier groups with antiship missiles. Picking up the quiet hum of one in busy coastal waters is "like trying to identify a single car engine in the din of a city," says Rear Adm. Frank Drennan, commander of the Naval Mine and Anti-Submarine Warfare Command. So the Pentagon has commissioned SAIC, a Virginia-based contractor, to build a sub hunter that will use powerful sonar to track enemy subs for months.

"This is all part of our new guiding concept, this idea of strategic agility," says Welsh. "It's the ability to adapt and respond faster than our adversaries." Then he starts in about a proposed hypersonic aircraft that will travel more than 5,000 mph.

Intelligent Machines: The Marines Meet Their Robot Mule

THERE'S A ROBOT FOLLOWING MARINE PFC. MARCUS BEEDLE THROUGH THE WOODS AT FORT DEVENS, MASS. The quadruped, the Legged Squad Support System (LS3), does this by using flickering lasers and stereoscopic cameras in its head to fixate on a pattern of thick bands strapped to the Marine's backpack. The robot also traces the path Beedle takes by tracking a navigation device strapped to his right boot. LS3 can pick its own way through rough terrain or tramp directly in its master's footsteps. "Follow-the-leader is our bread and butter," says Kevin Blankespoor, VP of controls and autonomy at Boston Dynamics (purchased by Google in December 2013), the creator of LS3. Beedle and a squad of Marines from the 1st Battalion, 5th Regiment, are the first to take this robotic pack mule for squad-level testing. *Popular Mechanics* is the only media outlet on hand for this historic meet-up of Marine and machine. Officials from the Marine Corps Warfighting Lab, DARPA, and Boston Dynamics developed the LS3 to go where tracked and wheeled vehicles can't go—over rocks, up and down steep inclines, and through woods and swamps. The idea behind the LS3 is for Marines to use it to take supplies to their secured positions. LS3 is made for war zones, but it is not viewed as a weapon. It's just a mule, though some Marines would like to see that change. "We'd love a machine gun on it," says 1st Lt. Alex Hurran.

→ Gear of the Future Marine

1. Cargo Drones

Unmanned helicopters have proven themselves by supplying Marines in Afghanistan since 2011. This March, the Corps extended their tours of duty indefinitely. By that time the Lockheed Martin aircraft, called K-Max, had hauled 3 million pounds of cargo. More resupply and attack drones are expected in the future.

2. Laser Defenses

The Marines are researching a defensive laser that can be installed on a 4WD vehicle. "We're looking at a counter-unmanned aerial-system capability right now, but in the future we'll look at rockets, mortars, and artillery," says Maj. Gen. Robert Walsh, deputy commanding general of the Marine Corps combat development command. "We've left a gap [in these defenses] over the past 11 years. But we have to be able to counter them."

3. Portable Power

The Office of Naval Research is creating a hybrid solar, thermal, diesel, and JP-8 fuel system that can replace large, loud trailers that provide electricity to Marines in the field. The project's managers hope the system will consume 40 percent less fuel than current systems, operate much more quietly, and run on biofuels.

4. Railgun

The concept has been around for nearly a century: Use electromagnetism instead of explosive charges to fire artillery rounds. The U.S. Navy has been working on just such a weapon since 2005—a railgun prototype that's compact enough for a ship. Designed to support Marines during land strikes and harass enemy vessels from afar, the prototype harnesses a 32-megajoule jolt of electricity to launch a 23-pound shell that can destroy targets up to 110 miles away with kinetic energy alone. (One megajoule is the energy equivalent of a 1-ton vehicle moving at 100 mph.) Here's how the railgun works: High-voltage capacitor banks are connected to two copper rails—one positively charged ❶ and the other negatively charged ❷. To fire the weapon, a current pulses down the positively charged rail, across a conductive armature ❸ that cradles a shell❹, and up the negatively charged rail. The completed circuit generates powerful electromagnetic fields that propel the armature and shell along the rails at tremendous velocity. At the end of the rails, the shell detaches ❺ from the armature and speeds to its target at more than 5,600 mph. The launch sequence takes 10 milliseconds. The Navy plans to start shipboard tests in 2016.

5. Minesweeping Drone

The autonomous Knifefish uses sidescan sonar to detect floating or buried mines. A lithium-ion battery powers the 19-foot-long craft on preprogrammed missions that can last up to 16 hours. Deployment is scheduled for 2017.

6. Unmanned Sub Hunter

Radar, lidar, and other sensors enable SAIC's Anti-Submarine Warfare Continuous Trail Unmanned Vessel to avoid ships while tracking quiet diesel-electric subs on 80-day missions that can span 3,800 miles. Prototypes are expected to hit the water in mid-2015.

7. Enhanced Combat Helmet

Kevlar and Twaron combat helmets, which are made of ballistic fibers, can withstand a direct hit from a 9-mm pistol round or even some bomb fragments. The Enhanced Combat Helmet, which the Army and Marine Corps began issuing late last year, is the first helmet capable of stopping a rifle round. It is made of ultrahighmolecular-weight polyethylene, a type of thermoplastic, and weighs about 3 pounds, the same as the other helmets.

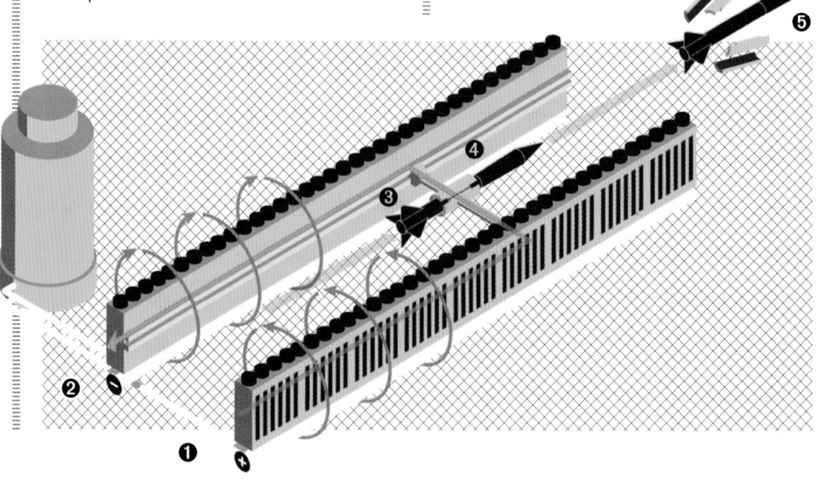

THESE WEAPONS, DRONES, AND BATTLEFIELD DEVICES—SOME RECENTLY DEPLOYED, OTHERS IN EARLY R&D—WILL HELP THE PENTAGON FIGHT MORE EFFECTIVELY WITH FEWER TROOPS.

8. Wearable Batteries

Gadgets can give U.S. troops an edge, but the penalty is the weight of all those batteries. Arotech developed the Soldier Wearable Integrated Power Equipment System to eliminate the need to bring along extra power packs. The system uses high-capacity zinc-air and lithiumion batteries to continuously charge a soldier's two-way radio, GPS unit, and other devices. Worn in a tactical vest, the thin, flexible power source extends missions and reduces battery weight by 30 percent.

9. Night-Vision Contact Lenses

Engineers at the University of Michigan built the first room temperature light detector with broadband sensitivity, from ultraviolet wavelengths through visible light to infrared. The photo sensor, which is made of two layers of transparent graphene, each a single atom thick, could eventually be integrated with circuitry into contact lenses. The wearable tech could eventually provide troops with thermal vision, a type of night vision for detecting heat signatures from vehicles, weapons, and troops.

10. Chemical Weapon Antidote

Chemical weapons remain a silent, deadly scourge. During an attack, when every second counts, troops could slap on this patch, which subcutaneously delivers an antidote via hundreds of microneedles.

11. Tendon-Assisted Rigid Exoskeleton

The Department of Veterans Affairs has reported a tenfold increase in musculoskeletal injuries from 2003 to 2009, not surprising given that troops in Iraq and Afghanistan carry packs that weigh 60 to 100 pounds. In response, cadets at West Point invented the Tendonassisted Rigid Exoskeleton (T-REX), a 3D-printed brace made of high-grade plastic and worn over combat boots. The latest prototype strings 100 ultrafine Flexinol wires between the heel and the ankle of the brace; the wires flex in response to electrical signals, contracting like a second set of muscles to boost leg strength. A pair of braces, which are powered by radio batteries the troops already carry in their packs, weighs just 1.5 pounds.

A TEXAS COMPANY CALLED TRACKINGPOINT HAS DEVELOPED A LASER-GUIDED, SEMIAUTOMATIC RIFLE WITH A NETWORKED TRACKING SCOPE THAT MEASURES DISTANCE, HUMIDITY, AND 16 OTHER BALLISTIC VARIABLES. THE PAYOFF IS HIGH FIRST-SHOT PRECISION ON TARGETS UP TO 750 YARDS AWAY. SHOWN HERE ON THE AR 762, THE SCOPE ALSO STREAMS LIVE VIDEO FROM ITS HEADS UP DISPLAY TO SMARTPHONES AND TABLETS. OTHER FIGHTERS COULD USE THE VIDEO TO ACQUIRE TARGETS; COMMANDERS COULD USE IT TO ASSESS BATTLEFIELDS IN REAL TIME. THE ARMY BEGAN TESTING A MILITARY VERSION OF THE RIFLE IN MAY 2014.

EDITOR'S NOTE

As combat operations in Afghanistan began winding down, the K-Max cargo drone was put to work closing bases in the Helmland province before finally being relegated to storage. It is estimated that since 2011, the vehicles hauled 4.5 million pounds of cargo around combat zones. —DHW

X-PLANE RISING

→ The future of naval aviation may not belong to pilots. The Pentagon is developing a cutting-edge warplane that can operate from a carrier without an aviator or remote controls. Will unmanned aircraft end the era of **Top Gun**?

BY **JOE PAPPALARDO**

ADDITIONAL REPORTING BY SHARON WEINBERGER

This article was published in *Popular Mechanics* in 2011.

F**OR ENSIGN KYNDRA CHITWOOD, LEARNING HOW TO** fly blind is just part of becoming a U.S. naval aviator. The 23-year-old strides across the flight line at Naval Air Station Whiting Field. Around her, dozens of orange and white T-34 Mentor training aircraft are lined up, fuselages gleaming in the Florida sun. Pairs of student-trainer teams in flightsuits are making their way to and from the aircraft. A formation flight of two T-34s cruises overhead, wingtip to wingtip.

X-47B

ALTITUDE: ***40,000 FT***

AUTONOMOUS ***AERIAL REFUELING PROBE & DROGUE***

WINGSPAN: ***62.1 FT***

LENGTH: ***38.2 FT***

POWERPLANT: ***PRATT & WHITNEY F100-PW -220U***

CREW: ***NONE***

TWO ***WEAPON BAYS: 4,500-LB CAPACITY***

NORTHROP ***GRUMMAN IS MAKING THE NAVY'S NEWEST EXPERIMENTAL WARPLANE, A CARRIER READY UNMANNED AIRCRAFT CALLED THE X-47B, SHOWN HERE DURING STRUCTURAL STRESS TESTS AT THE COMPANY'S FACILITIES IN PALMDALE, CALIFORNIA.***

Chitwood is readying herself for a tense afternoon in the air. A typical five-day week at Whiting, 25 miles northeast of Pensacola in the Florida panhandle, features daily flying tests. Each is a high-pressure evaluation. Today, Chitwood must prove she can pilot the T-34 using instruments alone.

Just after takeoff, she will pull a nylon hood across half of the cockpit canopy to block her vision, then steer the single-engine prop plane through landing approaches at several civilian airfields. She will be asked to demonstrate her ability to use GPS signals, UHF transponders on the ground, and radio commands from radar operators.

Within 40 minutes of suiting up, Chitwood and her instructor are cleared for takeoff. The aircraft buzzes down the runway and sails into the sky, banking toward Marianna Municipal Airport, 120 miles east. She would normally handle takeoff on her own, but the instrument tests require that she sit in the T-34's back seat, so her instructor guides the aircraft into the air.

The clouds are so thick she can't see the horizon, so there's no need to pull the hood closed. Once her instructor removes his hands from the controls, the day's aerial evaluation will begin.

Primary flight instruction here is the start of a process that makes U.S. naval pilots the best in the world. Every student in the class is graded on a bell curve, and those who score highest usually get first choice of flight assignments.

"I DIDN'T EVEN REALIZE, UNTIL I REALLY STARTED DIGGING, HOW ADVANCED SOME OF THE THINGS THAT WE'RE DOING ARE," HE SAYS. "THIS IS, NO KIDDING, MAKING AN AIR VEHICLE THAT'S AUTONOMOUS AND AS SELF-SUFFICIENT AS A NAVAL AVIATOR."

—CAPT. JAIME ENGDAHL

These aviators are jockeying against one another for seats in cockpits, but the generation of naval pilots after Chitwood may be grounded by new competition: robots. The Navy is aggressively researching the use of unmanned aerial vehicles (UAVs) for jobs now performed by people. By the time a pilot like Chitwood retires in 2030, assuming a full career, there could be a lot fewer aircraft for her replacement to fly.

TRIKE FIGHTERS ARE THE TEETH OF AN AIRCRAFT carrier, protecting the ship from aerial threats and attacking targets on the ground. The idea that UAVs can perform these missions is heretical to officials who say that a carrier deck is too complex for an unmanned aircraft. But the orthodoxy is changing, and two naval aircrafts now vie for future dominance.

In one corner, there's the Navy's marquee future warplane, the F-35C Lightning II. A product of the Joint Strike Fighter program, the F-35C will be the Navy's first stealth aircraft. At $133 million per airplane, it is the most expensive defense program in the world.

The JSF program started in 1996. Five years later, Lockheed Martin beat Boeing for the multibillion-dollar contract. The program is producing three F-35 variants: one for the Navy, another for the Air Force and a short-takeoff and vertical-landing version for the Marine Corps. Each aircraft is now nearing the end of a tortured development—years late and tens of billions of dollars over budget.

ENSIGN ***KYNDRA CHITWOOD DOESN'T BELIEVE HER CAREER AS A NAVY PILOT IS THREATENED BY UNMANNED AIRCRAFT. "IT'S HARD TO INTEGRATE THEM INTO THE FLEET," SHE SAYS.***

In the opposing corner is the X-47B, an experimental airplane with something to prove. Even though it's just a demonstrator, it has folding wings that enable it to fit inside a carrier's hangar, twin weapons bays, and the ability to fly at high subsonic speeds. The program started in 2000 as one of two $2 million concept studies, but the UAV is no longer a line-item underdog.

The X-47B is a testbed supporting the $2.5 billion Unmanned Carrier Launched Airborne Surveillance and Strike program. UCLASS may result in an unmanned aircraft that can perform the same missions as the F-35C but would stay in the air longer and be harder to spot on radar.

When he was selected to head the X-47B program in late 2010, Capt. Jaime Engdahl thought the UAVs would be remotely piloted onto the deck with a joystick. "I didn't even realize, until I really started digging, how advanced some of the things that we're doing are," he says. "This is, no kidding, making an air vehicle that's autonomous and as self-sufficient as a naval aviator."

A few years ago, predictions that the F-35C would be the last piloted Navy fighter seemed overly dramatic. But as the X-47B progresses, the prediction is more realistic. The Navy will not deploy the F-35C on a carrier until late 2016. The Pentagon plans to integrate strike UAVs into the fleet by 2018.

Despite stalwart support from Navy brass, politicians are considering trimming the F-35C and other JSF variants. In 2010, the National Commission on Fiscal Responsibility and Reform (Bowles-Simpson Commission) recommended halving orders of F-35Cs.

There are signs that the budget crisis may change the landscape within the Navy. Aviation Week obtained a 2011 memo from U.S. Navy Undersecretary Robert Work asking Navy brass to seek alternatives to the F-35C. (He also asked the Marine Corps to examine the impact of eliminating the problematic F-35B.) The response will be ready for the 2013 budget.

The Navy's need for the F-35C and the new UAVs is based on the emergence of fresh threats. Any future strike aircraft needs to be stealthy—advanced radar and antiaircraft missiles make strike and surveillance missions dangerous.

EDITOR'S NOTE

Indeed, the Navy's 2015 five-year budget plan reduced orders of the F-35C from 69 to 36. In November of 2014, however, a series of historic test flights demonstrated the fighter could successfully take off and land on an aircraft carrier at sea. —DHW

→ Future Flight: The Contenders

HIGH-PERFORMANCE STRIKE AIRPLANES ARE THE A-LIST CELEBRITIES OF THE NAVY. These warplanes must be tough enough to handle the stress of catapult launches, sophisticated enough to evade electronic attack, rugged enough to withstand salty ocean air, nimble enough to dogfight a MiG, and deadly enough to eliminate ground targets with precision. Four U.S. Navy warplanes fit these criteria. All are vying for primacy as the Navy reshapes its future.

Phantom Ray

MANUFACTURER: Boeing
SELLING POINT: The company completed its first experimental stealth UAV, the X-45A, in 2000 and sporadically improved the prototype.
STATUS: In 2011, Boeing paid for Phantom Ray test flights, positioning itself for a pending Navy strike-aircraft competition.

F-35C Lightning II

MANUFACTURER: Lockheed Martin
SELLING POINT: The F-35C, the carrier variant of the Joint Strike Fighter, is the most sophisticated warplane ever built. Pilots will be able to see threats from 360 degrees and access targeting data from other aircraft and ground sensors.
STATUS: After years of delays, the F-35C may deploy in 2016. Congress is considering cutting the number of planes it orders.

X-47B

MANUFACTURER: Northrop Grumman
SELLING POINT: The Air Force and Navy created the predecessor to this demonstrator, and that version has already proven it can drop weapons and autonomously fly preplanned missions.
STATUS: The UAV's first carrier landing was planned for 2013. Automated aerial refueling tests will soon follow the landing milestone.

F/A-18E/ F Super Hornet

MANUFACTURER: Boeing
SELLING POINT: The Super Hornet, fielded in 1999, has made a career filling the gaps left by failed fighter programs. Although it's not a stealth aircraft, designers shaped the airplane to reduce its radar cross section.
STATUS: As the F-35C faced delays, the Navy ordered more F/A-18s.

EDITOR'S NOTE

Based on a successor to the X-45A (the predictably named X-45C) the latest Phantom Ray prototype was unveiled in May 2013. With a 50-foot wingspan, the sleek unmanned dogfighter was touted as having a slew of advanced capabilities, including intelligence gathering, in-flight refueling, and the ability to suppress enemy air defenses.

On July 10th, 2013, the X-37B Unmanned Combat Air System successfully performed a carrier-based arrested landing onboard the USS George H.W. Bush, marking the first time a tailless, unmanned autonomous craft has landed on a modern aircraft carrier.

Barely a year later in June 2014, a massive engine failure occurred on an Air Force F-35 jet. In response, the U.S. Navy issued a grounding order for all F-35 B-model and C-model fighter jets. For now, the future of this expensive warplane is murky.

The tried-and-true F/A-18E/F Super Hornet continues to serve in the United States Navy and as the primary fighter jet of the Royal Australian Air Force (RAAF), available and ready as the F-35 program falters. —DHW

There are other threats that will likely force carriers to operate at greater distances from targets. China is fielding submarines that can lurk in the shallows, where side-scan sonar is less effective. These quiet subs are armed with sea-skimming missiles that can slip past a ship's defenses. The Russians build and sell sophisticated warplanes that can venture from air bases on land to swarm a carrier and its escorts with air-to-ship missiles.

The emergence of these close-to-shore threats is bad news for the Navy because the reach of its strike aircraft has been decreasing since the 1980s. The F-35C's 640-nautical-mile combat radius will reverse the trend, but UAVs such as the X-47B easily double that distance.

And there's another metric to consider: the amount of time a stealth warplane can linger over a target. The military calls this attribute persistence. A manned aircraft's persistence is limited by the endurance of the human onboard. A weaponized UAV, on the other hand, can track a target for dozens of hours in protected airspace and then drop a precision weapon on command. "My thinking is that [UAV adoption] is too damn slow," Adm. Gary Roughead, chief of naval operations, said during a speech in August. "We've got to have a sense of urgency about getting this stuff out there."

N JULY 2011, LT. JEREMY DEBONS MADE AVIATION history by doing nothing at all. DeBons, call sign Silas, flew a one-of-a-kind F/A-18D through the vast Atlantic test range off the coast of Virginia. His destination: the aircraft carrier USS Eisenhower. The F/A-18D and the carrier were loaded with a slew of sensors that enabled the warplane to land on the carrier's deck without any piloting from the cockpit or remote operation by the ship's crew.

The F/A-18D was a surrogate for the X-47B demonstration aircraft. The goal of the summer tests was to prove that the brains of a robot can guide a warplane to a carrier's pitching deck. If this demonstration program succeeds, the Navy can proceed with the UCLASS procurement. After the X-47B becomes carrier-ready, it will then autonomously rendezvous with aerial refueling tankers.

Future unmanned aircraft will not be flown by crews via joystick, as Air Force personnel currently operate MQ-1B Predators. Once launched, a UAV following a prearranged mission plan will use onboard sensors to avoid other aircraft and dodge enemy attacks. It will also identify targets in the air and on the ground and track them without direct command. (These UAVs will contact controllers for permission to release weapons.)

Landing on an aircraft carrier is one of the most difficult feats of aviation, requiring a clever mesh of man and machine. The Navy is building the X-47B's landing capability on technology pilots use today, the Precision Approach Landing System (PALS), which uses SPN- 46 radar to locate an aircraft in relation to a carrier.

To perform the occasional hands-off landing, such as when weather obscures the ship, an F/A-18 pilot can couple the plane's autopilot with PALS data. But the PALS radar covers only the rear of the carrier and is limited by the number of aircraft it can track simultaneously. These deficiencies make it unsuitable for controlling UAVs that are approaching to land.

Instead, the Navy's robotic landing system relies on precise GPS coordinates to obtain 360-degree coverage and automate navigation. The airplane calculates the appropriate flight paths around the ship as the carrier supplies the vessel's speed, the sea state, and other data.

The concept behind the X-47B is to replace pilots–but not a ship's crew. As with other carrier borne aircraft, the final approach of the unmanned vehicle would be monitored by the personnel onboard. Officers on the flight deck, the air traffic controller seated under the deck, and officers peering from windows in the carrier's island all play a part in guiding the UAV.

Pilots most often talk with the ship by radio, but for a UAV, verbal communication is replaced by digital commands. The carrier's air traffic controllers do the same job as an ATC crew in a civilian airport tower–if the runway were constantly changing location, the aircraft loaded with live weapons, and the process designed to produce as few electromagnetic emissions as possible to avoid detection.

"There's no benefit in changing the way we do deck handling," says Adam Anderson, who heads the carrier integration portion of the X-47B program at Patuxent River, Maryland. "We're looking at ways to make the least amount of impact. We want a paradigm shift in the number of missions the aircraft can do, but to have no shift in the way it lands."

DeBons's test flights on the Eisenhower went perfectly–the F/A-18D glided across the deck with its nose angled up until a hook on its tail snagged a cable stretched across the deck and jerked the aircraft to a halt. Still, the Navy test pilot says his hands never strayed too far from the controls.

"It wasn't anything new, perspective-wise, in the cockpit," DeBons told reporters after the flight. "But being a new system, as any test pilot will say, we're always on guard."

HE X-47B'S NEXT BIG MILESTONE IS A series of aircraft carrier launches and landings in 2013. That is the same year that the F-35C will operate for the first time on a carrier, making its initial sea trials.

A technology demonstrator is a long way from a production-ready warplane like the F-35C. Even the most gung-ho Navy official extolling UAVs also expresses support for the Lightning II. "As rapidly as we want to engage with the unmanned systems on carriers, we are also moving forward with an incredible capability in the Joint Strike Fighter," Roughead says. "We've got to get that aircraft." The pilot is the F-35C's main limitation, but the human being in the cockpit may also be its salvation. The F-35C is designed to accommodate and enhance the most powerful processor available—the human brain. The aircraft's external sensors are patched directly into the pilot's helmet, allowing him to see 360 degrees by synthesizing data from the sensors, including six infrared cameras and radar. In short, there has never been a better airplane for picking targets and seeing threats.

"Target recognition often involves the generation and interpretation of high-resolution images," Owen Cote Jr., the associate director of MIT's Security Studies Program, wrote in a recent report. "At some point in the future it may become possible to automate that process, but today, and for a number of years, target recognition will require people to interpret the images . . . It is difficult to imagine automating this."

Others caution putting too much faith in stealthy UAVs as a solution to every tactical and budgetary problem at the Pentagon. "If you canceled the F-35, you would have to do something else," says Douglas Barrie, military aerospace senior fellow at the International Institute for Strategic Studies. "Do you kick off a whole new development program with all the inherent risks and costs that you've just gone through with the F-35?"

It appears that the Navy's near future will focus on human-robot teaming. The F-35C and UCLASS will operate from carriers simultaneously and fly missions next to each other in tomorrow's conflicts. "No platform fights alone anymore," says Edward Timperlake, a former analyst of emerging technology for the office of the secretary of defense. "It's a synergy between manned and unmanned."

But as time goes on, UAVs will become more capable and the ratio of manned to unmanned missions could dramatically shift in favor of the machines. "The era of manned airplanes should be seen as over," the Brookings Institution's Michael O'Hanlon says, echoing other military analysts.

Few young pilots see a threat looming. After all, the number of people the Navy needs to fly is holding steady. The students at Naval Air Station Whiting Field say their careers remain unclouded by robotic competition. "I don't see anyone concerned about the community shrinking," says Lt. J.G. Bobby Lennon, who finished training at Whiting in August and will fly an MH-60S Knighthawk helicopter. "Maybe this will affect the jet guys more."

But the Navy's effort goes beyond the X-47B. The service is investing $8 billion over the next five years in a family of unmanned UAVs. For example, the Navy plans to retire fixed-wing EP-3 Aries signals intelligence reconnaissance planes and replace them with unmanned aircraft by 2020. The Navy is already operating unmanned reconnaissance helicopters—one was shot down during combat operations in Libya—and it is constructing larger ones to carry cargo and arming others with air-to-surface missiles.

Chitwood, who aced her instrument flight examination over Marianna Municipal and Tallahassee Regional airports, is also unfazed by the emergence of UAVs. "I'm not really worried," she says. "It's hard to integrate them into the fleet. Maybe when I'm close to being done flying I'll see the effect."

It doesn't seem fair to bring up the Navy's unmanned programs when face to face with young aviators. That's policy talk, and every new Navy pilot is solely focused on the work. For them, earning the right to do the job is the only tomorrow that matters.

→ Machines on the Move: Balancing Act

THE BIGGEST CHALLENGE FOR WALKING ROBOTS USED TO BE STAYING UPRIGHT. Now, Boston Dynamics has two four-year DARPA contracts to make its Petman android prototype as agile as its human creators. Most robots utilize static stability, always keeping their center of mass directly over the support provided by their legs. But to move quickly, the new robot, called Atlas, will practice dynamic stability—instead of reacting to motion, the robot will anticipate how the step changes its balance and compensate by swinging its arms or adjusting its feet. Boston Dynamics will use Atlas and four-legged robots, such as BigDog (a 2006 *Popular Mechanics*-Breakthrough Award winner) and Cheetah (built to run at least 30 mph), to study robot mobility. Future dynamically stable rescue robots could climb through rubble or crawl into tight spaces to assess building damage or search for survivors.

AL-QAIDA *FROGMEN BEWARE: THE U.S. DEPARTMENT OF HOMELAND SECURITY WILL BEGIN FIELD-TESTING AN UNDERWATER ROBOT AT AN UNDISCLOSED PORT THIS YEAR. THE BIOSWIMMER, IN DEVELOPMENT BY BOSTON ENGINEERING CORPORATION, MIMICS A TUNA'S FORM AND MOVEMENT TO SCOUT CONSTRAINED SPACES IN HARBORS, NUCLEAR-PLANT WATER INTAKES, AND EVEN THE BILGE TANKS OF SHIPS.*

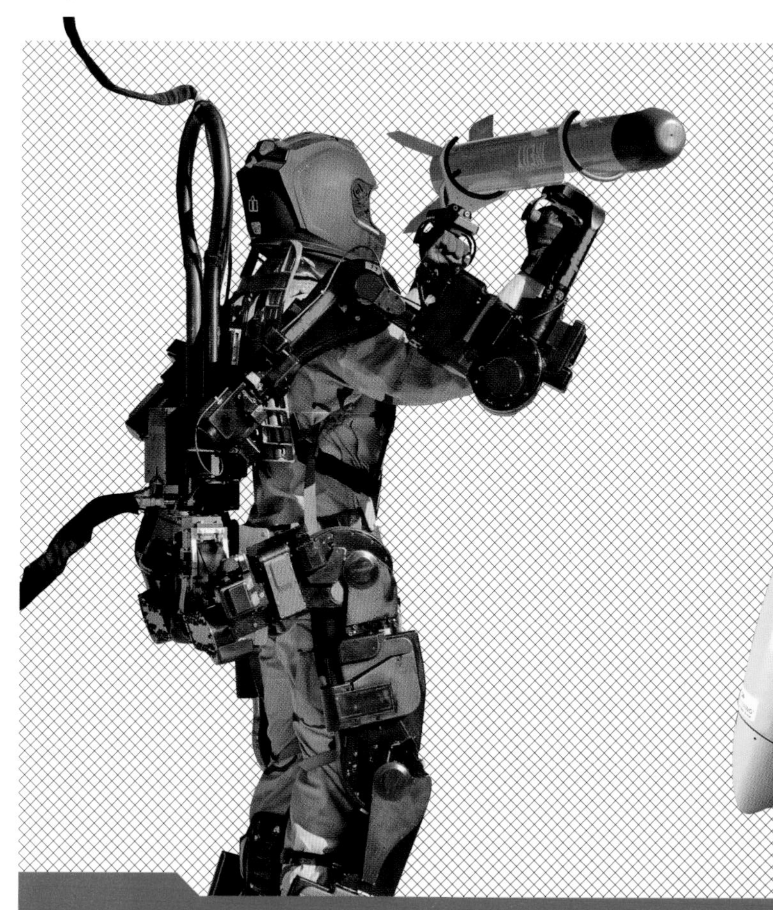

EDITOR'S NOTE

Since Google purchased Boston Dynamics in December of 2013, they have moved to shut down all DARPA contracts. In other words, watch for this technology to appear in your living room before you see it on the battlefield. —DHW

RESEARCHERS *PLAN TO OPTIMIZE THE PETMAN ANDROID (LEFT) TO NAVIGATE ROUGH TERRAIN WITH UNPRECEDENTED AGILITY. THE ROBOT MAY ULTIMATELY BE ABLE TO SWING FROM ONE HANDHOLD TO THE NEXT LIKE A HUMAN.*

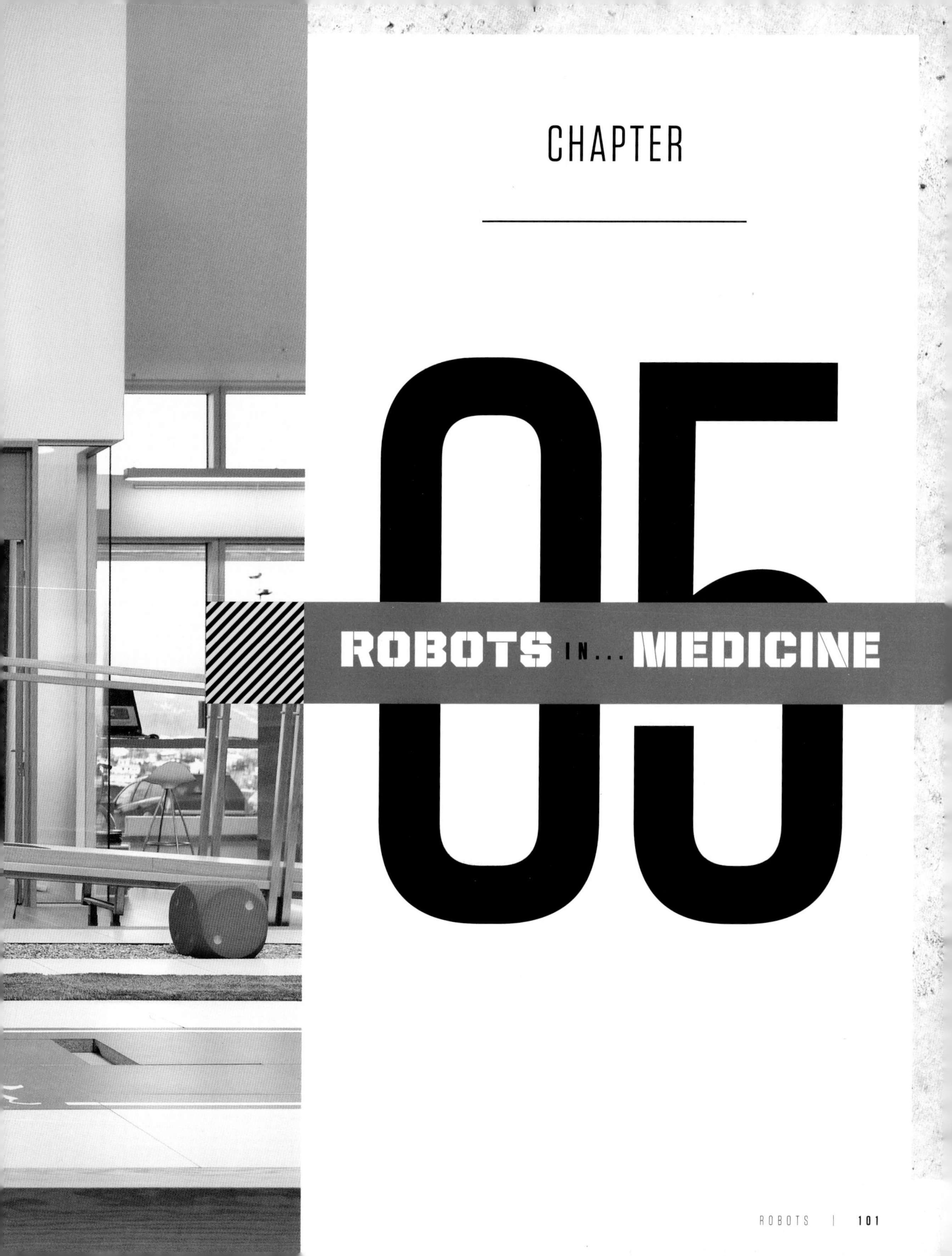

CHAPTER

05

ROBOTS IN... MEDICINE

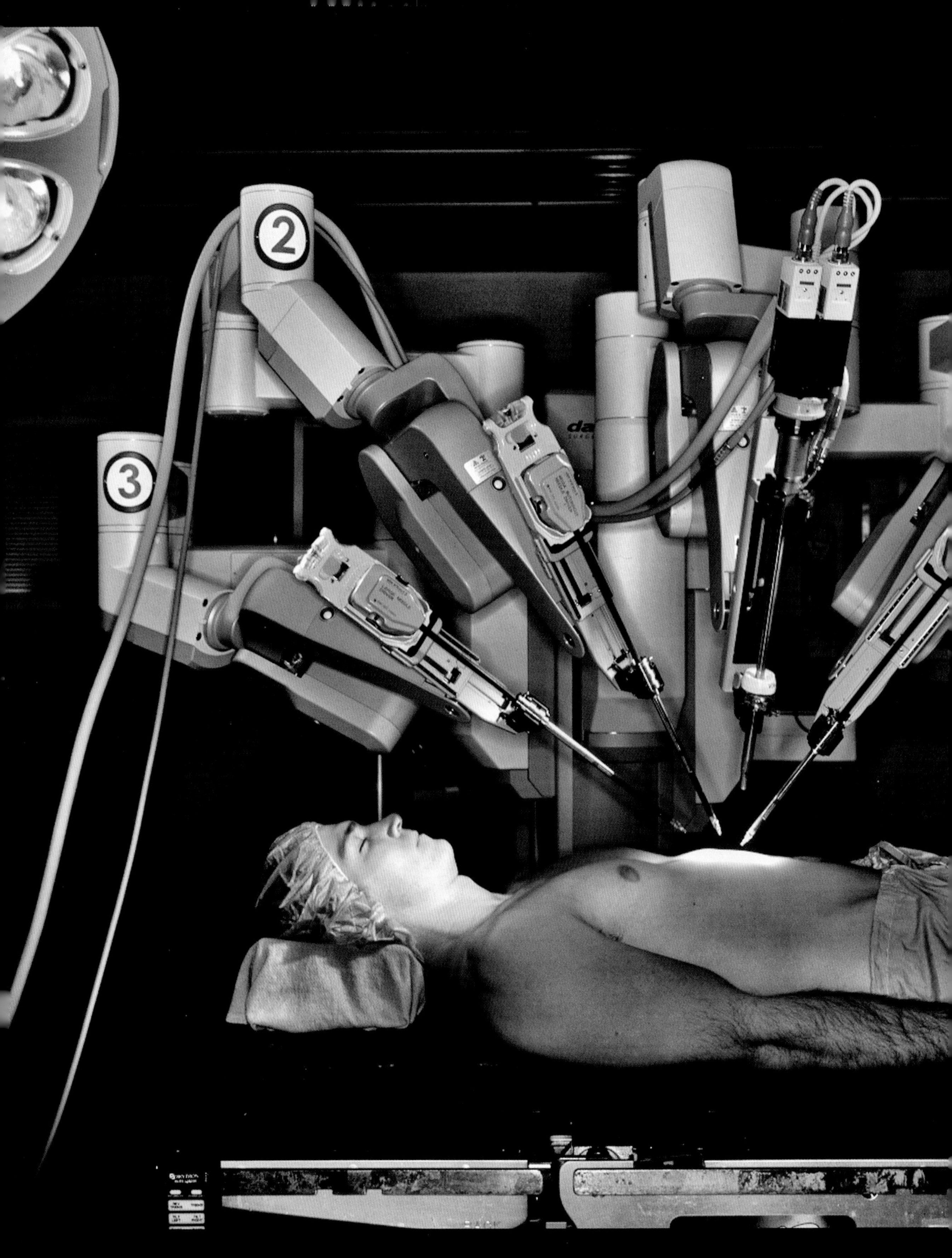
2
3

THIS WON'T HURT A BIT

➔ By 10 am, the robot has already carved out an entire human organ.

BY **ERIK SOFGE**

This article was published in Popular Mechanics *in 2013.*

HE PATIENT'S GALLBLADDER, RIDDLED WITH INFECTION, is gone, and in a darkened operating room at Boston's Beth Israel Deaconess Medical Center, the machine is going back for seconds.

Its next target is a tumor, buried in that same woman's kidney. So the bot remains perched; its arms fanned out over her like some great, hulking insect. Those arms twitch and wriggle, and inside the sleeping woman's abdomen its tiny, dexterous manipulators slice and burn through fatty connective tissue, maneuvering around veins, arteries, and nerves, clearing a path to the cancer with a level of precision no human could ever hope to muster.

The robot, called the da Vinci Si Surgical System, isn't performing a partial nephrectomy under its own volition. It is being controlled by two human surgeons seated at a pair of consoles in a corner of the OR. They take turns commanding the machine, using an array of foot switches and hand controls to snip or sear through tissue. Their movements are replicated and interpreted by the robot a dozen or so feet away—the da Vinci smooths out the control signals, eliminating physiological tremors from the doctors. The robot enhances, even as it obeys.

This isn't some experimental test or limited pilot program—it's simply how surgery is done today. The da Vinci system, which first reached hospitals 14 years ago, has become the most common surgical robot on the planet, with almost 2,500 units worldwide performing over 200,000 procedures per year. And while the bot was initially used in urologic surgery, it's since racked up so many procedures that the system's manipulators now touch nearly every internal organ. Instead of encircling the patient in a tight, huddled

THE *DA VINCI SI SURGICAL SYSTEM MACHINE LOOKS MENACING, BUT THE ROBOT REMAINS UNDER THE CONTROL OF A HUMAN.*

pack, the nurses, technicians, and surgeons in a da Vinci operating room each have a discrete station in a different part of the OR, and all the specialists are staring at screens showing the same intimate, zoomed-in contours of the patient's insides. Over the past decade, robots have reshuffled a work environment that had taken more than a millennium to perfect and have transformed an entire profession.

"THE WORD ROBOT IS VERY MISLEADING. THEY THINK I PUSH A BUTTON AND THEN WALK OUT OF THE ROOM."

—ANDREW WAGNER, BETH ISRAEL DEACONESS MEDICAL CENTER

"The surgeon's role is totally different now," says Andrew Wagner, director of minimally invasive urologic surgery at Beth Israel and the attending surgeon for the procedure I watched. "A couple times a month, a patient will say, 'Let me see your hands.' They want to see how steady they are. And I do it. I play along. What I don't tell them is that it really doesn't matter. Motion scaling removes that part of the job."

According to Wagner and the growing number of surgeons and hospital administrators who've seen the da Vinci in action, the benefits of robot surgery come from the ability to synthesize human intelligence with machine-assisted precision. Like the majority of other surgical robots, the system specializes in minimally invasive procedures. Its slim, cable-driven manipulators fit into relatively small incisions, fully or partially removing organs more nimbly than the poles used in traditional laparoscopic surgery, and with less trauma than open surgery. There's less violence done to the patient, which not only turns potentially massive scars into minor ones, but can also reduce the rate of complications and recovery time. Whereas open prostatectomies typically require a three-day hospital stay, a patient could go home within 24 hours of a robotic procedure, if not the same day.

"We've moved from doing prostate and kidney surgery all open to 90 percent robotic," Wagner says. While many of his patients are excited by the prospect of being operated on by a robot, others are terrified. "The word robot is very misleading. They think I push a button and then walk out of the room," Wagner says.

Eight years ago, the problem with the Renaissance surgical bot wasn't underperformance— in fact, it was doing too much. Still in development at the time, the machine was designed to guide a drill to specific positions along a patient's spine and, when given the go-ahead by the attending surgeon, bore away. But during preliminary testing, Israel-based Mazor Robotics found that orthopedic surgeons wanted the robots to have less autonomy. They were fine with letting the robots aim, but they wanted to pull the trigger and retain that tactile feedback of the bit spinning through bone. So, Mazor CEO Ori Hadomi says, the company demoted its creation and removed the automatic drill.

The Renaissance bot is a relative newcomer to the OR, with just 18 systems in the U.S. so far, but it's not alone. The wide range of partially autonomous surgical robots currently deployed includes the Lasik laser eye surgery system, Mako Surgical's joint resurfacing RIO system, and the CyberKnife system, which doesn't physically cut into patients but hits them with precisely targeted doses of radiation. These are true robots—machines that are given specific orders but whose programming determines how to execute those tasks.

The CyberKnife, for example, isn't simply a medical-grade particle accelerator with good aim. Its algorithms allow it to hit a moving target, adjusting its beam to accommodate a patient's breathing or other involuntary movements. Someone who might have needed a six-week course of radiation from a traditional linear accelerator can spend as little as a week with the CyberKnife. And since the machine is hitting bull's-eyes every session with an emitter that conforms to the irregular shape of a given organ, there are fewer side effects and less crippling discomfort.

The CyberKnife might be saving patients who, with traditional treatments, would have been written off. Initial results from an ongoing study at the University of Pittsburgh indicate that CyberKnife treatments can turn pancreatic cancer patients, often considered inoperable, into viable surgical candidates. Dwight Heron, a professor at the university and an oncologist, says, "This is something we've never seen before." By shrinking the tumors with precisely targeted radiation before subjects go under the actual knife, surgeons can potentially remove one of the most lethal cancers. "We're talking about patients where surgeons say, 'We can't take it out. Maybe you'll live a year.' Now they'll live potentially five years and may even be cured," Heron says.

ITH DEMAND FOR SURGICAL SYSTEMS ON the rise, academic researchers are developing a second wave of medical robots, systems with even greater degrees of autonomy. In 2010 the Duke University Ultrasound Transducer Group demonstrated, using turkey breasts, a robot that could perform completely unassisted biopsies. Beyond testing its core capabilities on two women diagnosed with breast cancer, the machine hasn't made it to broader human trials. But the group's director, Stephen Smith, believes the technology could have a huge impact in the developing world. "We envision a mobile van with a mammography unit, a 3D scanner, a robot, and a PC with AI software," Smith says. "One technologist would do everything."

At Carnegie Mellon University, robotics professor Howie Choset is still waiting for the FDA to clear the first commercial application of his snake-shaped surgical robot, a system called the Flex. The Massachusetts-based Medrobotics, which Choset co-founded in 2005, is presenting the flexible robot as a natural fit for ear, nose, and throat procedures. Choset wants to do more, hoping to develop similar systems that not only compete with the da Vinci, but further redefine what surgery is and who can perform it. "This will disseminate medical care," Choset says. "When we operated with surgical-snake robots on pigs, we had nonsurgeons doing the jobs that surgeons used to do." Users pilot the system with a simple joystick. And while the operator would need to know enough about human anatomy to avoid getting lost, the incisions would be considerably smaller and in areas that heal more readily– in orifices such as the mouth, for example.

Choset sees the first application for such a system on the battlefield, where a field medic and an intuitively controlled snake-bot could be someone's sole hope for survival. "The number of hours it takes to master surgery with the da Vinci is about the same as it would take for an open procedure–10,000 hours," Choset says. Autonomy, along with the winding, flexible nature of a snake-bot, promises to exponentially reduce training time. "If you can play a video game, you can drive our robot," he says. It's a hopeful take of the future of surgery, but back at Beth Israel, it's coming apart at the seams.

For a relatively unhurried 90 minutes or so, Wagner and his partner have cleared away fatty tissue and located the urine and blood pathways to avoid, laying the groundwork for the tumor's removal. Now the artery feeding into the kidney is clamped, and the clock has started. With no blood pumping in, the organ is effectively dying. Wagner has 20 minutes to not only get the cancer out, but also cauterize and suture the kidney's worst wounds before the clamp comes off. Around 15 minutes in–or 5 minutes to go–blood arcs across the screen, just barely missing the camera.

"Whoa. Hello," Wagner says. "We have a pumper. Watch out, Steve. Don't let it hit me in the face." While Wagner's camera bobs and weaves, urology chief resident Steve Eyre uses an old-fashioned, nonrobotic pole to nudge this prodigious bleeder into a safer trajectory.

As the clamp comes off and the scrambling subsides, it sinks in. In this particular operating room and during this particular procedure, there's no place for autonomy or lightly trained technicians. You need experts to look at that tumor. You need Steve to reposition tissue and organs. You need multiple people making complex, sometimes urgent, decisions. And even if human judgment could be somehow distilled into code, our flesh is too unpredictable.

"Every patient is different," says Catherine Mohr, director of medical research at Intuitive Surgical. In theory, a self-guided version of the da Vinci could be loaded with a general map of the region and the ability to find and remove the prostate. Half of the time the results might be suitable. But the other half, when the nerves don't line up, you'd wind up with patients who were impotent, incontinent, or both.

It's a problem of imaging, really. Hard tissue, such as bone, is easier to scan. But soft tissue is still a puzzle box, with vessels and nerves and plumbing that show up as ghosts or best guesses before surgery starts. So while systems like the Renaissance and Mako's RIO can follow a concrete game plan established ahead of time using clear X-rays, MRIs, and ultrasounds–and reconfirmed during the procedure–the only guarantee with soft tissue is that it will be messy.

Without the ability to autonomously navigate soft tissue, self-guided bots can't be trusted to go diving into organs. They still have roles to play, though, in more contained missions. When a robot like the RIO resurfaces a patient's joint to better interface with a new implant, it's functioning like a kind of surgical CNC machine, manufacturing a component within the body. The Renaissance system makes a compelling case for specific bursts of autonomy–in spinal procedures where a deviation of 2 millimeters can mean repeating the surgery or possibly even paralysis, a machine that's accurate to within 1 millimeter is a clear benefit.

The future of surgery is not a linear path, with the da Vinci and its master-

→ The Robotic Spectrum

Mako Surgical RIO

The RIO is a single, superaccurate robotic arm on wheels. It can be fitted with various tools to resurface degraded or diseased joints or to position reconstructive implants and is currently used on hips and knees. Despite steady sales of the RIO, Mako Surgical has struggled financially. Before the robot can move to other parts of the body, Mako will have to solidify its role in lower-body joint surgery.

Mazor Robotics Renaissance

The Renaissance, a robotic guidance system for spinal procedures, positions the surgeon's drill during each step, combining pre- op scans with realtime X-rays to provide pinpoint accuracy. Currently used on the spine, it's FDA-cleared for cranial surgery, though specific procedures haven't been described in detail.

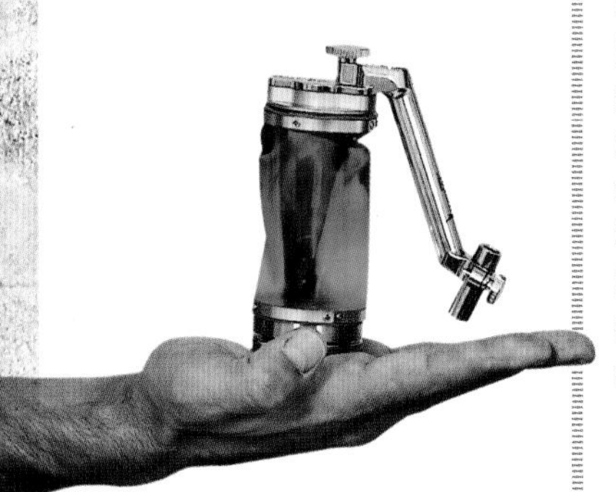

Accuray CyberKnife

While traditional radiation therapy bathes entire regions of a patient's body with high-energy rays, the CyberKnife's particle accelerator fires precise, pulsed beams at specific organs. Rather than expanding use to new areas of the body (the CyberKnife is currently used on breast tissue, the female reproductive system, gastrointestinal tract, head and neck, intracranial, kidney, liver, lung, pancreas, prostate, and spine), the engineering team wants to develop more focused beams to treat patients whose recovery previously might have been ruled out.

Intuitive Surgical da Vinci

The world's most common surgical bot, the da Vinci system allows remote control of a 3D camera and up to three instrumented arms that enter the body through tiny (1- to 2-cm) incisions. It's currently used on adrenal glands, colon, heart, gallbladder, kidney, prostate, spleen, stomach, throat, and the female reproductive system. No new regions are planned, so any in-development techniques or procedures are essentially variations on current ones– or, at the very least, will involve the same organs.

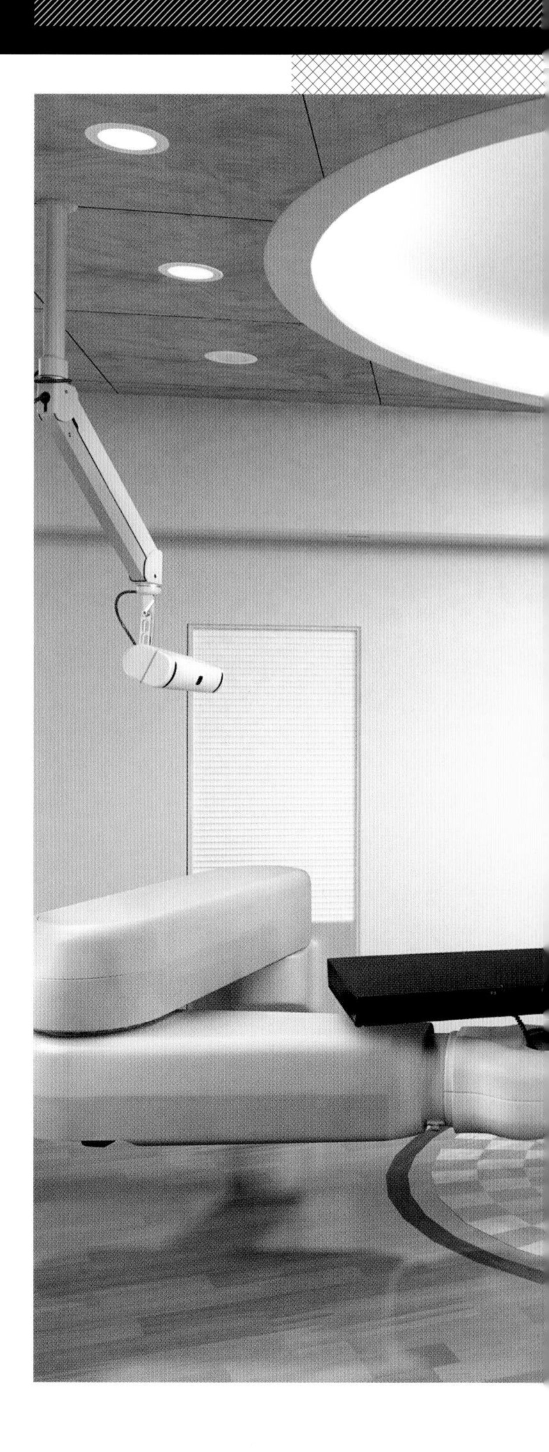

ROBOTIC SURGICAL SYSTEMS, SUCH AS THE RENAISSANCE, NOW OPERATE ON ALMOST EVERY MAJOR INTERNAL ORGAN. AND THEY'RE COMING TO AN OPERATING ROOM NEAR YOU.

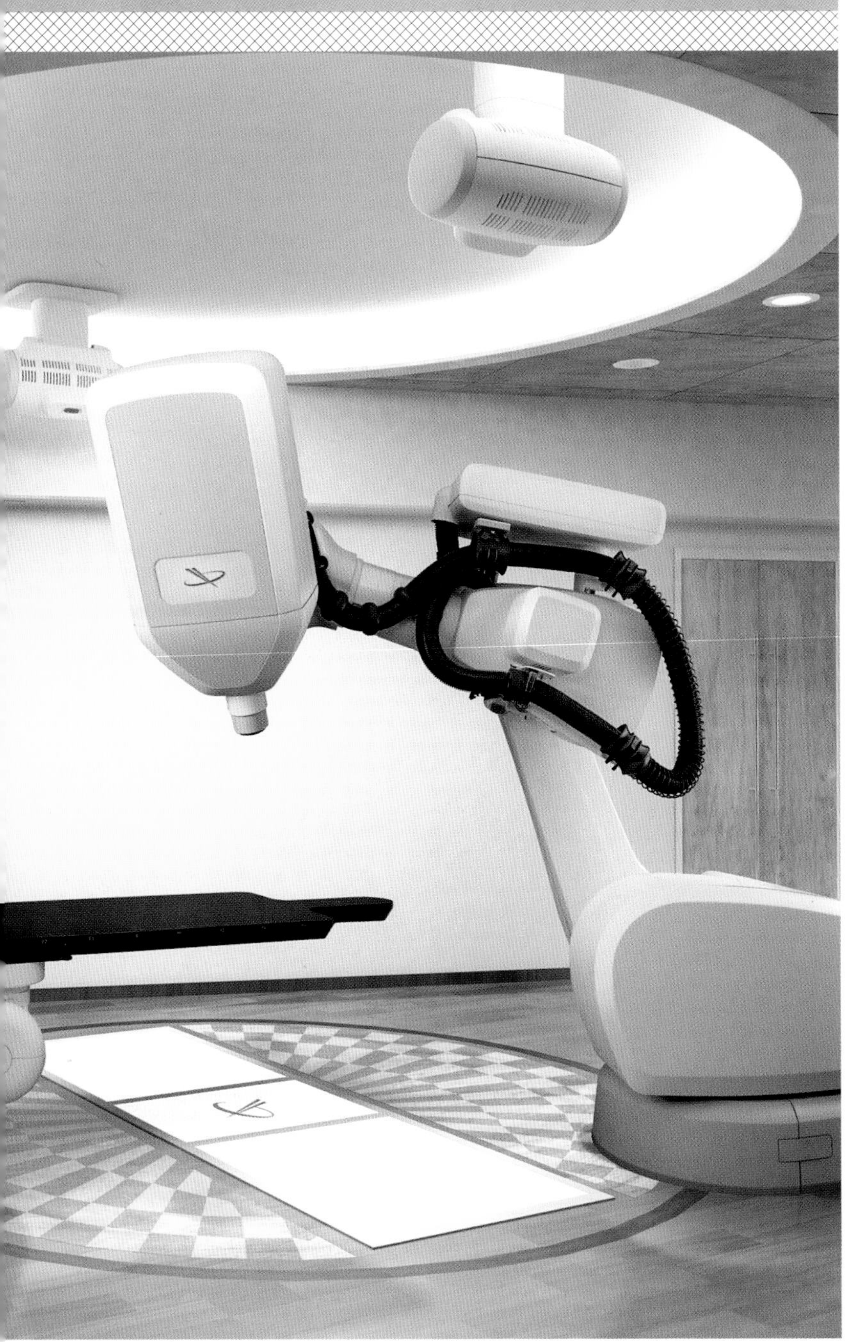

slave control signal laying the foundation for more autonomous systems, such as the CyberKnife and Renaissance. Automation and teleoperation are simply two halves of a collective, robotic solution. And barring some quantum leap in artificial intelligence, there'll always be a time and place for human cognition.

What's certain is that the era of human hands inside another body is drawing to a close. To Wagner, for procedures he tackles on a weekly basis, the human-only approach is already obsolete. "I've done something like 600 prostate surgeries. I've never done one open," Wagner says. But he's the one doing the surgery, not a machine alone. Still, he says, "In my mind there's no need, really, to learn that operation the traditional way. We've proved that it's just better for everyone with the robot."

EDITOR'S NOTES

At publication, the da Vinci surgical system is being used in over 400,000 surgeries a year. The fourth generation da Vinci Xi surgical system—with better vision systems, smaller appendages, and even laser targeting—received FDA clearance for use in April 2014.

Mazor Robotics Renaissance continues to make progress, selling new units in the United States and Asia for use in complex spine surgery. The Stryker Corporation acquired MAKO Surgical Corporation in January 2014, and intends to apply the robotic arm technology to joint reconstruction throughout the body. —DHW

Medical Innovations 2015–2050

BIOBOLT

Engineers continue to refine the next generation of this wireless device, tested in monkeys, which uses skin as a medium to transfer neural data to a bionic limb. The real breakthrough will be when the process provides sensory feedback. The hardest part will be writing the algorithms, says Euisik Yoon, principal investigator at the University of Michigan. "No one really knows what to say to the brain. Not yet."

2013

GEN 3 ARM

DEKA's Gen 3 arm has 10 degrees of freedom and six preset grasp patterns. The company, led by Dean Kamen, plans to launch the arm with a wireless foot controller, though it could also accept commands from electrodes in the user's nerves or brain.

2013

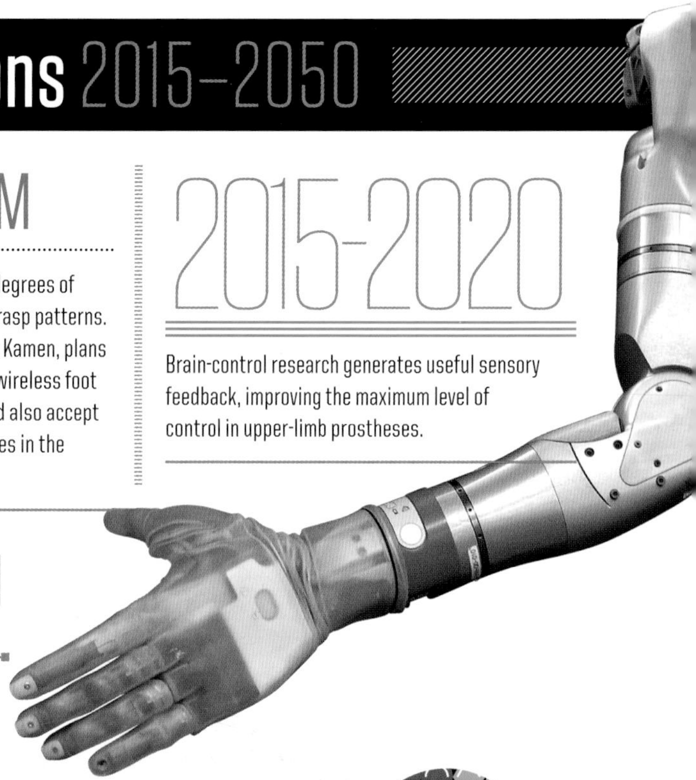

2015-2020

Brain-control research generates useful sensory feedback, improving the maximum level of control in upper-limb prostheses.

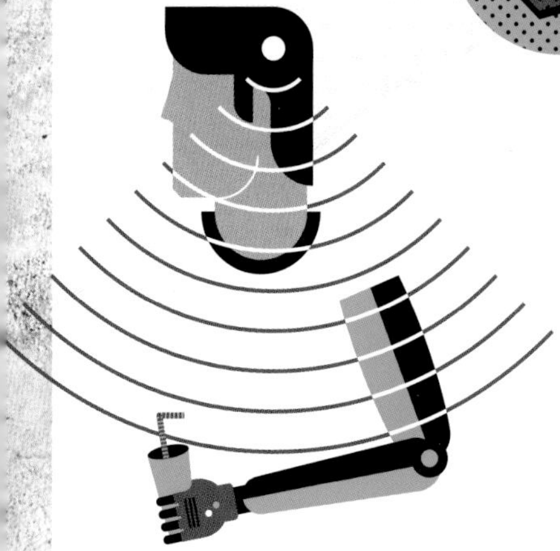

2013-2015

After successful human trials, wireless Brain-computer interface (BCI) systems begin replacing wired systems, particularly for prosthesis control.

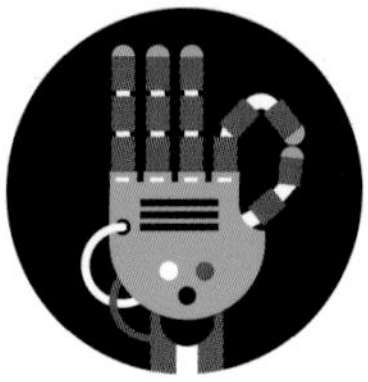

2016

MODULAR PROSTHETIC LIMB

Pioneered at Johns Hopkins University Applied Physics Laboratory, the Modular Prosthetic Limb (MPL) has 26 articulating joints. The first commercial version will likely be controlled via electrodes on or under the skin. But arms manipulated by sensors directly on the nerves and by brain implants are in clinical trials now.

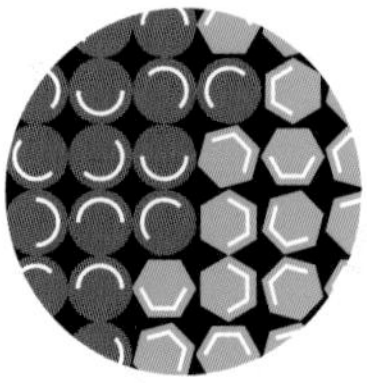

2017

ARTIFICIAL WHITE BLOOD CELLS

These tiny plastic "smart particles," called leuko-polymersomes mimic white blood cells, finding and adhering to infected cells. They can also be programmed to target specific cells and perform a variety of tasks, such as delivering a cancer drug. The University of Pennsylvania's Dan Hammer says the invention of more organically compatible parts could accelerate FDA approval.

PRINTED BONES

Washington State University scientists created a ceramic powder that they can use to print artificial bone scaffolds, employing a specially configured inkjet printer and CAD software. The scaffolding dissolves as natural bone grows around it. Lab-printed load-bearing bones, such as hips and knees, still remain at least 10 years off. "You have to really respect Mother Nature," chemist Susmita Bose says. "She is very, very hard to mimic."

2022

ARTIFICIAL KIDNEY

An artificial kidney built at the University of California, San Francisco, replaces all the functions of the real thing in an implantable device about the size of a baseball. A series of nano-drilled silicon screens filter toxins out of the blood without the need for pumps or outside power. A bioreactor filled with specially engineered kidney cells performs other important renal functions, such as maintaining electrolyte balance.

2022

BIONIC LENS

University of Washington researchers have been working toward a wirelessly powered on-eye display, aiming for a commercial version in 10 years. With micro-optics embedded in the lens, it could not only provide night vision or projecting data into the line of sight, but also restore some degree of lost vision. "For macular degeneration, the incoming image can be modified to be spread over the still-functional parts of a retina," project lead Babak Parviz says.

2022

ARTIFICIAL RED BLOOD CELLS

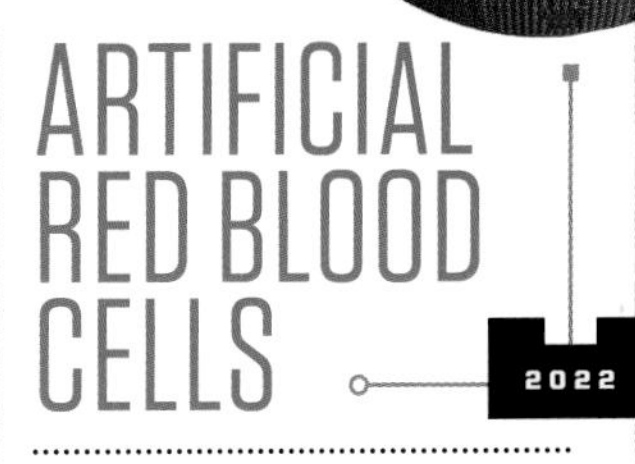

2022

Also called respirocytes, these robotic blood cells consist of about 18 billion atoms—most of which are carbon—in a diamond structure that's assembled one atom at a time. They could carry hundreds of times more oxygen than natural red blood cells, allowing people to hold their breath for hours or sprint at Olympics-level speeds for 15 minutes at a time. Robert Freitas of the Institute for Molecular Manufacturing hopes to have the first nanofactory up in 20 years.

2025

Exoskeletons that augment specific capabilities, such as strength or speed, hit the market, along with restorative ones that respond to wireless BCI.

2027

LAB-GROWN LIVER

Anthony Atala's team at the Wake Forest Institute for Regenerative Medicine has already implanted lab-grown bladders. Next up: more complex organs, such as the liver, grown from a patient's own cells on a scaffold. While a full implant is likely 15 years off, early work could provide injections of healthy cells or grafts of lab-grown tissue to repair damaged organs.

2050

Prostheses are now obsolete, replaced by reconstructive technology.

→ Bionic limbs with machine intelligence can now sense their environment and predict a user's intentions. Smarter, stronger, speedier robotics will someday enhance the powers of the able-bodied, too.

BY **ERIK SOFGE**

This article was published in *Popular Mechanics* in 2012.

HERE AREN'T ENOUGH BIONIC MEN ON THE planet to produce a proper stereotype. Even so, David the Farmer seems atypical. Ruddy, red-haired, and impossibly cheerful when he meets us on the gravel path outside his workshop. What I was expecting—a grizzled retiree limping stiffly through his daily chores—bears no resemblance to this 30-something mechanic climbing down from a massive tractor without hesitation, weaving between ATVs and scattered engine parts, and moving from task to task with no evidence that he's part machine. After a few minutes, there are clues, though: He always turns on his right leg, and his pants

AN *ENGINEER AT ÖSSUR RUNS DIAGNOSTICS ON THE RHEO KNEE'S PRIMARY MOTORIZED JOINT. PART OF THE SYMBIONIC LEG, THIS PROSTHESIS CONTAINS ANGLE AND FORCE SENSORS THAT SEND REAL-TIME DATA TO AN ONBOARD COMPUTER.*

DAVID *INGVASON IS A BIONIC LEG'S WORST NIGHTMARE—THE FARMBASED MECHANIC EXPOSES THE WORLD'S FIRST INTEGRATED BIONIC LIMB TO ICELAND'S MOST GRUELING CONDITIONS. ON RARE OCCASIONS, HE'S BEEN KNOWN TO RECHARGE IT FROM A TRACTOR'S DC JACK.*

gather around his left ankle, hinting at a limb that's slightly skeletal and decidedly nonbiological.

Today, that left leg will be replaced. That's why engineers from Össur, one of the world's largest prosthesis-makers, drove an hour west from the company's headquarters in Reykjavík, Iceland, to the farm where David Ingvason lives and works. David the Farmer—the nickname they've given their star prosthesis tester, though he is actually employed as a full-time, on-site mechanic—is one of a limited pool of amputees fitted with the Symbionic Leg: an artificial knee, ankle, and foot that are integrated into a single bionic limb.

On the farmland and surrounding terrain, in tall grass, and on moss-sprayed plains of volcanic rock, Ingvason regularly destroys his leg. He fouls the motors in muck and sludge, burns them out through unremitting use, and generally grinds one of the most sophisticated auto-adaptive devices on the planet, each one worth more than some sedans, into an inert, cybernetic paperweight. According to Össur's new technology search manager, Magnús Oddsson, all Ingvason has to do is call and they'll hand-deliver a new limb. More often, he swings by Reykjavík himself wearing a backup leg and asking for a repair or replacement. Whatever David the Farmer wants, he gets—the punishment he metes out to his leg, and the data that result, are simply too useful.

Össur began selling the Symbionic model as the world's first commercially available bionic leg last fall. It represents a significant shift in prostheses. The traditional stand-ins for lost limbs and senses are now being imbued with machine intelligence. Ingvason's leg is, in fact, a robot, with sensors that detect its environment and gauge his intentions and processors that determine the angle of his carbon-fiber foot as it swings forward. The same approach is being applied to prosthetic arms, in which complex algorithms determine how hard to grasp a water bottle or when to absorb the impact of a fall. Vision- and hearing-based prostheses bypass faulty organs and receptors entirely, processing and translating raw sensor data into signals that the brain can interpret. All of these bionic systems actively adapt to their users, restoring the body by serving it.

Take, for example, one of the most common prosthesis failures. A mechanical knee typically goes rigid as the heel lands, supporting the user's weight and then unlocking when pressure is applied to the toe. If that toe contact comes too early, the leg collapses under its owner. The Symbionic Leg isn't so easily fooled. Force sensors and accelerometers keep track of the leg's position relative to the environment and the user. Onboard processors analyze this input at a rate of 1,000 times per second, deciding how best to respond—when to release tension and when to maintain it.

Since the leg knows where it is throughout the stride, achieving a rudimentary form of proprioception, it takes more than a stubbed toe to trigger a loose knee. If the prosthesis still somehow misreads the situation, the initial lurch of the user falling should activate its stumble-recovery mode. Like antilock brakes for the leg, the actuators slow to a halt, and magnetically controlled fluid in the knee will become more viscous, creating resistance as the entire system strains to keep the person from crumpling or toppling.

The result, Ingvason says, is that he rarely falls, or at least no more than someone with two biological legs. He can drive ATVs, hike across glaciers, and even ride a horse while herding sheep. "I don't have to think about it," he says. Before he went bionic, Ingvason fell constantly. "With the old knee, it was every day, often more than once in a day," he says. Now I'm walking on uneven ground and high grass and sand and mud and everything."

→ Electromechanical Medical Miracle: Indego Exoskeleton

FORTY-TWO-YEAR-OLD MICHAEL GORE OF WHITEVILLE, N.C., REMEMBERS WHEN HE COULDN'T WAIT TO PLOP INTO A CHAIR AFTER A 12-HOUR SHIFT AT THE VINYL-FENCING PLANT WHERE HE USED TO WORK. "Now it's the opposite," he says. "I can't wait to stand up." For the past decade, Gore has been in a wheelchair, paralyzed from the waist down after a workplace fall that wrecked his spinal cord. But when he's strapped into the Indego, a powered exoskeleton developed by a team led by Vanderbilt University engineer Michael Goldfarb and his former graduate student Ryan Farris, Gore can stand, walk, and even climb stairs.

The Indego fits tightly around the torso and extends down to the ankles. Powerful, battery-operated electric motors drive hip and knee joints. The user operates the system by leaning forward to stand up or walk and leaning back to sit down. Standing and walking, even for a few hours a week, addresses health problems that plague paraplegics, including loss of bone density and poor blood circulation. To commercialize the exoskeleton, Goldfarb partnered with Cleveland-based Parker Hannifin, which hired Farris to join the Indego team. The company hopes to make the Indego available in rehab clinics, with a consumer model to follow.

EDITOR'S NOTE

As of publication, the ReWalk exoskeleton has been approved for at home use, while the Ekso is approved for use only in rehabilitation centers. The Indego still awaits FDA approval for use at home or in rehabilitation centers. —DHW

MICHAEL *GOLDFARB (LEFT), RYAN FARRIS, AND INDEGO EXOSKELETON, PHOTOGRAPHED BY DANIEL SHEA AUG. 7, 2013, AT SHEPHERD CENTER, ATLANTA.*

Ingvason's newly delivered limb is another Symbionic Leg, loaded with upgraded software that will allow the knee and the ankle to communicate with each other. Össur plans to develop this feature over the coming years, establishing what Oddsson calls networked intelligence. After putting it on, Ingvason limps awkwardly at first across dirt and gravel, past the rusting hulks of trucks and cars. Within a few minutes, the robot has calibrated itself.

With Ingvason's pant leg hitched up, it's impossible not to watch the limb in action. It's harsh and alien. The gray polymer shell, which partially conceals aircraft-grade aluminum, seems too skinny to support his weight; the ankle, too delicate for the 10,000 newtons of force it was built to withstand. But the leg is nimble and so quick to react, it's as though he were born with it.

HE GOAL OF CURRENT BIONIC RESEARCH is to recover what was taken. In Ingvason's case, it's the leg he lost nearly 12 years ago, when he stopped to help a couple whose car had broken down in the rain. While he was working, another vehicle slipped off the wet road and plowed into him. Others' losses include arms torn off by industrial accidents or improvised bombs, and senses dulled or snuffed out by disease. Despite decades of study, the prostheses developed to replace lost functions have been at once ingenious and disappointing.

Most prosthetic devices create their own health problems. Purely mechanical legs use a complex system of gears and analog triggers to allow people to walk, but users must hike up one hip with each step to keep the artificial toe from scraping the ground. Powered prosthetic arms tend to be locked in place during walking—and that dead weight throws off the user's balance and posture. Roughly 70 percent of amputees develop back and joint problems, and experts suggest that such "co-morbidities" force those who might be obese or in chronic pain to become even less mobile and less healthy, ultimately shortening their lives.

The answer, for now, is in the algorithms. Össur's Symbionic Leg eliminates hip hiking through a simple robotic twitch: The toe actuates upward during each step, performing what's called dorsiflexion. Other algorithms are more sophisticated, interpreting a torrent of sensor data as specific types of terrain. If the foot lands at a higher elevation, with the knee bent, the leg assumes the presence of stairs and adjusts accordingly. If the toe tips up on contact and the heel dips down, the artificial intelligence (AI) suspects a slope and shifts the angle and resistance to assist in climbing.

The new generation of prosthetic arms has a different set of software challenges and solutions. DEKA, the research firm founded by inventor (and 2009 *Popular Mechanics* Breakthrough Award winner) Dean Kamen, is developing the third generation of its bionic limb, known internally as Gen 3. It's backed by DARPA's Revolutionizing Prosthetics program—a $100 million effort to create devices that are roughly equivalent in function to biological arms. Now awaiting FDA approval, Gen 3 has 10 degrees of freedom (typical motorized arms have only two or three) and a range of algorithms that mimic the precise control of its flesh-and-blood counterpart. By moving his or her foot, which operates a wireless controller, the user can engage various preset grasping pat-

THE ***MODULAR PROSTHETIC LIMB LOOKS LIKE AN ADVANCED BIONIC ARM. REALLY, IT'S A SWARM OF ROBOTS: EACH DETACHABLE, MOTORIZED SEGMENT HOUSES A PROCESSOR***

TO ***REENGINEER ORGANS SUCH AS THE BLADDER (ABOVE), SCIENTISTS AT THE WAKE FOREST INSTITUTE FOR REGENERATIVE MEDICINE POPULATE A BIODEGRADABLE SCAFFOLD WITH A PATIENT'S OWN CELLS. LED BY 2006*** **POPULAR MECHANICS** ***BREAKTHROUGH AWARD WINNER ANTHONY ATALA, THE TEAM IS NOW BUILDING MORE COMPLEX ORGANS, LIKE THE LIVER AND THE HEART.***

→ Fine-Tuned Prostheses

ENGINEERS CAN REPLICATE MISSING BODY PARTS FOR AMPUTEES, BUT RE-CREATING THE ABILITIES OF LOST LIMBS IS A MORE COMPLEX FEAT. ADVANCES IN ENGINEERING ARE MAKING PROSTHESES MORE FUNCTIONAL, ENABLING USERS TO HAVE RICHER LIVES.

FOOT FEEDBACK

Leg amputees often develop gait abnormalities because they can't feel what their prosthesisis are doing as they walk. A system developed at UCLA collects data from pressure sensors on the sole of a foot; the sensors trigger dime-size balloons encircling the thigh, which inflate to inform users when and how to correct their stride.

RECLAIMING CAREERS

Hand prostheses are too clumsy to allow carpenters who've lost fingers to keep working. Biomedical engineer Michael Morley, founder of ProSolutions, developed a prosthesis that restores the ability to swing a hammer, turn a screwdriver, and pull a handsaw.

"TASTING" SIGHT

Wicab's BrainPort gives sight to the blind—using their tongues. A pair of sunglasses converts digital images into electronic signals that are delivered via wire to a tongue strip. In lab tests, users with training could navigate hallways and distinguish between letters by the strength and pattern of the pulses "painted" on their tongues.

WALKING NATURALLY

Researchers at the Rehabilitation Institute of Chicago have created a prosthetic leg that anticipates its wearer's intentions. The limb uses the electrical signals from the amputee's flexing muscles to adjust its motion. For example, the leg would know when to bend the knee and when to point its toes up to climb a stair.

terns. Previous upper-limb models have used foot switches but with nowhere near the number of grip options, nor the machine intelligence and force sensors that guide artificial fingers and determine how much power should drive them. "The results have been incredible," says Stewart Coulter, the Gen 3 project manager. "The other day, one of our testers was eating with chopsticks, doing a better job than I could." The second arm funded by the Revolutionizing Prosthetics program, the Modular Prosthetic Limb (MPL), developed at Johns Hopkins University, may lead to what many believe is the endgame for bionics: direct neural control. By embedding electrodes into a subject's existing nerves, or going through the skull and implanting them directly onto his or her cortex, researchers have been able to turn thoughts into action. In a study conducted in 2010 at the University of Pittsburgh, a quadriplegic pressed the MPL's hand against his girlfriend's. Through trial and error, processors are taught to decrypt a user's thoughts and recognize a growing list of intentions. "The system's smart. It has to be," says Michael McLoughlin, Revolutionary Prosthetics' project manager at Johns Hopkins. "The algorithms interpret what the patient is trying to do, then do it."

The MPL, in other words, isn't truly mind-controlled. The electrodes deliver orders, but it's the arm that decides how to carry them out. Or, rather, it's the network of machines—each jointed segment and finger with its own processor—that makes up the arm. In the same ways, the state of the art in powered prostheses is stranger than science fiction: a swarm of bots that obey the human mind, either through cables that snake out of the skull or by taking their best collective guess at those thoughts. Stranger still, this is just the beginning.

TIP MY FOOT UPWARD. THE BIONIC foot that's bolted to the side of the table does the same. I press my toes into the floor and the prosthesis pivots downward into empty space. It's mirroring what I do, responding to the vibrations in my calf muscles, which are picked up by silicone-embedded microphones strapped to my bare leg. The system isn't detecting the full-blown tremors of muscle activity but a set of lower-level, initial rumblings that begin when the subject first intends to move.

Unlike the tests run in Össur's Gait Lab, where users wearing sensor rigs climb ramps and cross gravel and sand,

→ Robotic Touch

MUCH OF THE FINE MOTOR CONTROL IN THE HUMAN HAND RELIES UPON ITS ABILITY TO SENSE TEMPERATURE, VIBRATION, AND PRESSURE. "If your fingers are numb from the cold, your hands are almost useless," says University of Southern California biomedical engineering professor Gerald Loeb. Loeb and former graduate student Jeremy Fishel, cofounders of a company called SynTouch, have developed bionic fingers that give machines human-like touch. Using a flexible polymer skin and a variety of sensors, the robot fingers were able to correctly identify test materials 95 percent of the time by touching them, outperforming blindfolded human subjects. BioTac sensors could eventually lead to prosthetic arms with sensory neural feedback as well as to worker robots that operate safely in close contact with humans.

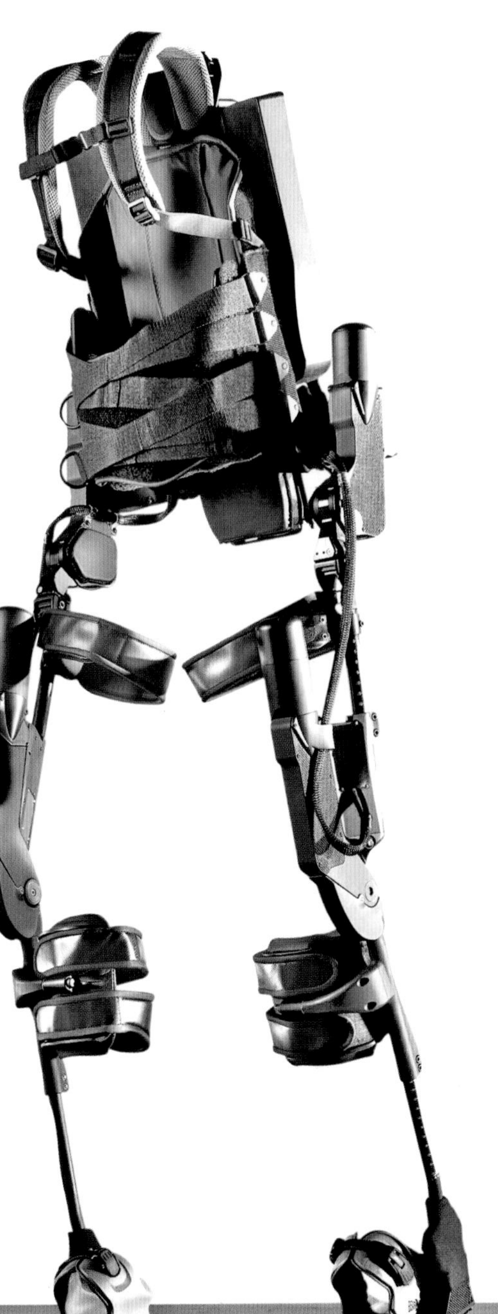

IN *FEBRUARY OF 2012, EKSO BIONICS' EXOSKELETON BECAME THE FIRST TO STEP OUT OF THE LAB AND INTO THE REAL WORLD. THIS MOTORIZED ROBOTIC LOWER-LIMB SYSTEM SUPPORTS THE FULL WEIGHT OF USERS WITH SPINAL CORD OR NEUROLOGICAL DAMAGE. ONBOARD BATTERIES SUPPLY 3 HOURS OF FULLY POWERED WALKING.*

this research is happening behind closed doors. It's part of the company's own quest to find an alternative to invasive neural control. "What we would like to do," Oddsson says, "is exactly what the user wants. And for that we need some kind of a brain-computer interface [BCI]." Like other efforts, it's a work in progress: The 125 milliseconds it takes for vibrations to be processed into action is still painfully slow compared with the near-instant reflexes of a biological limb. A foot muscle can respond to input within 40 milliseconds—faster than even the brain can deliver a response. But it's a technology worth pursuing.

In the long term, experts agree that while implanted interfaces could change the lives of millions of patients with amputations, spinal cord injuries, or neurological disorders, bionics that require major surgery will always be expensive, niche devices. For the millions suffering from debilitating strokes, or people with no serious disability but the money to pay for a wearable bionic system, a noninvasive BCI would change everything.

In other words, it's how we could reach that persistent fantasy of the able-bodied—true bionic augmentation. Even the most evasive experts I spoke with agreed that, while visions of superhuman amputees may be ridiculous, a combination of noninvasive BCIs and exoskeletons could turn decades of bionic research into a mainstream tool. California-based Ekso Bionics released the first commercially available exoskeleton in February; it's designed for patients with neurological or spinal cord damage. Billed as a "wearable robot," the system walks under its own power, currently via a remote control.

An advanced version translates shifts in balance and feedback from canes into a natural stride. "That's the right use of the technology now, as a medical device," Ekso spokeswoman Beverly Millson says. "But it's really a technology platform. It's the beginning of wearing your devices, for whatever purpose you might have."

At MIT's Media Lab, prosthetics pioneer (and 2005 *Popular Mechanics* Breakthrough Award winner) Hugh Herr is still in the early stages of bionic augmentation research, with a lower-limb exoskeleton system that cuts in half the forces associated with walking. Herr, who lost both legs below the knee as a teenager, understands the need for restoration. He's spun off a company, iWalk, to market his BiOM robotic ankle-and-foot system, now in clinical trials. Yet he says the endgame of his own work would be some kind of bionic vehicle that commuters might use to literally run to the office. Imagine sprinting 60 miles without breathing hard, across terrain that would stop an ATV in its tracks. "Something like the mountain bike will be completely laughable," Herr says. "We'll still have trucks to transport goods. At that point, though, driving alone across town in a metal box with four wheels would be just absurd."

Before leaving Össur, I coax a few specifics out of Oddsson. He's a scientist through and through. But he must have some sci-fi-tinged vision of what's beyond the Symbionic Leg. He tells me he wants to take what Össur has learned about the human body and the intricacies of gait—the tremendous forces and physics at work in a single step—and create something that hijacks the nervous system directly. "It's not an exoskeleton. 'Smart trousers' might be more accurate," Oddsson says a little sheepishly. The goal at first would be to help stroke victims. The device would stimulate the muscles, providing commands that the brain or damaged nerves can't. "We would use the actuators that are already there, the muscles, and simply provide a new central controller," he says.

Eventually, Oddsson says, prosthetics research will disappear, replaced by advanced reconstructive technology. By 2050, he ballparks, limbs will be recreated—printed, grown, who knows?—and all of the arcane, biomechanical secrets collected by companies like Össur will be harnessed to finally restore flesh and bone. It's a strange best-case scenario: that an industry will innovate itself out of existence, its research seeding other scientific fields while all of that sophisticated technology migrates toward devices that change the way millions of us—abled or disabled—live and work.

EDITOR'S NOTE

Medical exoskeletons continue to improve at a rapid pace. Wearing an FDA-approved ReWalk Robotic Exoskeleton, Claire Lomas completed the 2012 London Marathon in 16 days, despite being paralyzed from the chest down. —DHW

How It Works

BIOTAC FINGERTIPS, SUCH AS THE ONES ON THIS ROBOTIC HAND, EACH HAVE SENSORS TO DETECT HEAT, PRESSURE, AND VIBRATION. Taking a cue from human anatomy, BioTac digits even have textured "fingerprints" to improve sensitivity.

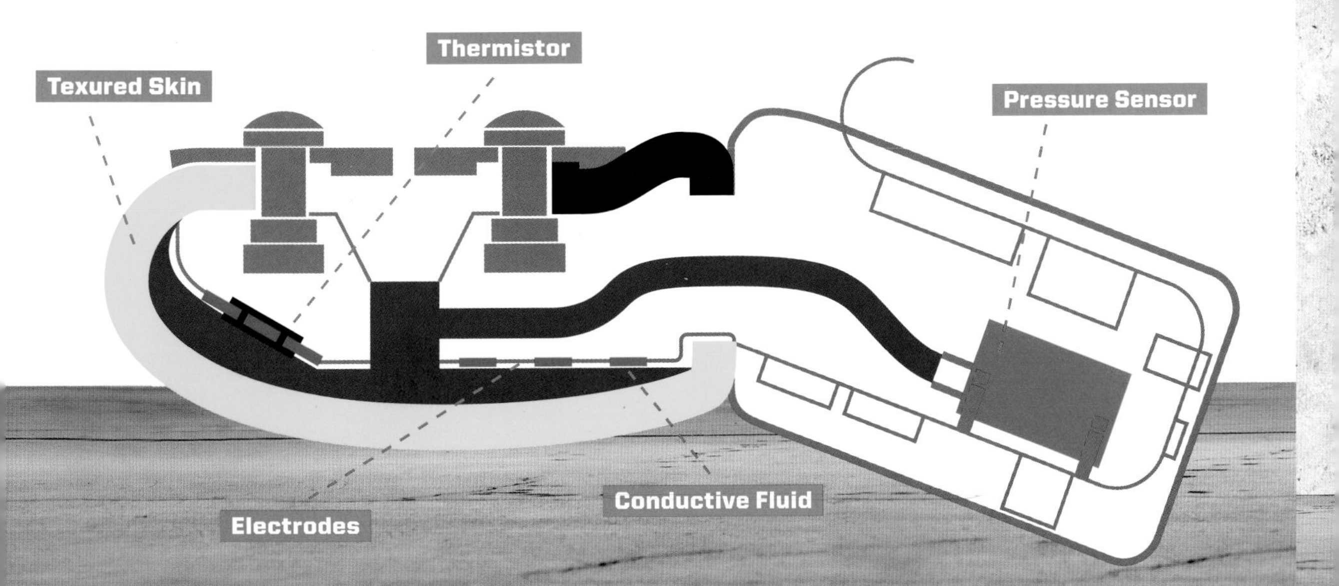

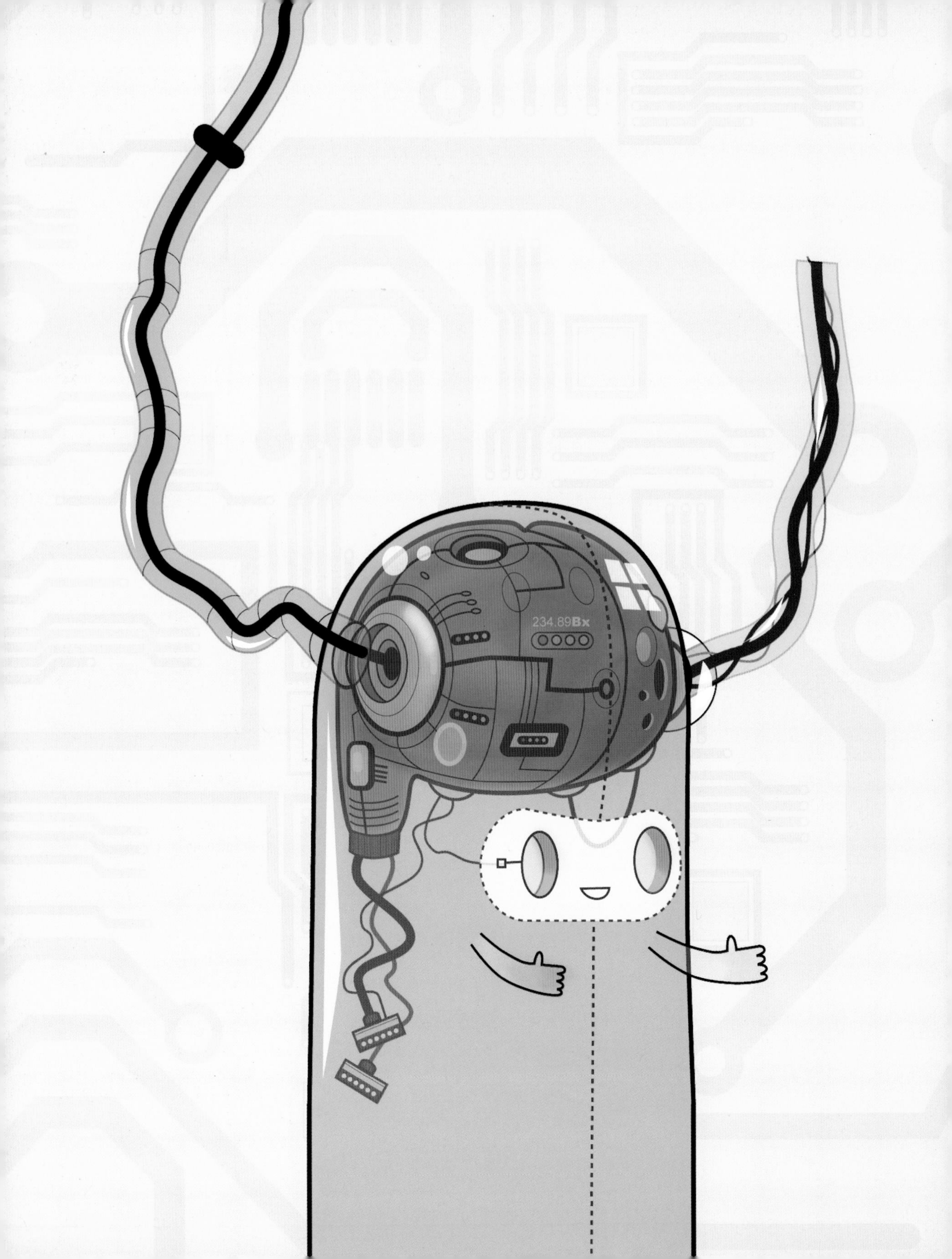
234.89Bx

VIRTUAL SMARTS

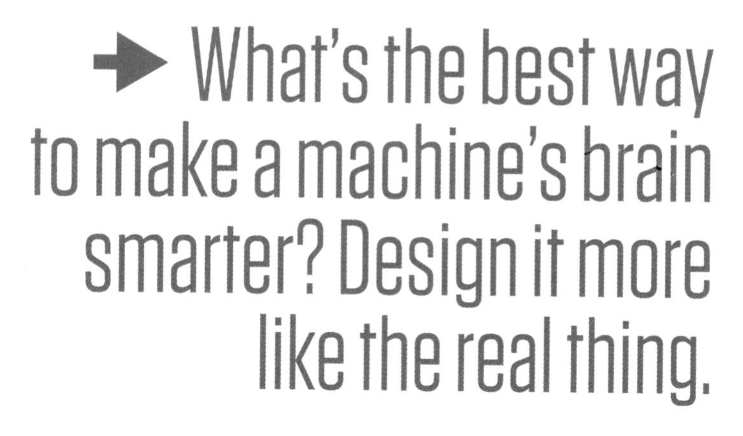

What's the best way to make a machine's brain smarter? Design it more like the real thing.

BY **DOUGLAS FOX**

This article was published in *Popular Mechanics* in 2010.

STEP INTO AN ACRE-SIZE ROOM AT Lawrence Livermore National Laboratory (LLNL) in California, and you come face to face with a giant. In the far corner, across an ocean of empty floor tiles, sits Dawn, a Blue Gene/P supercomputer–one of the fastest machines in the world. Its 147,456 processors fill 10 rows of slanted computer racks, woven together by miles of cable. Dawn's gentle name belies a voracious appetite. It devours a million watts of electricity (equal to 1,000 U.S. households) through cords as thick as a bouncer's wrists, racking up an annual power bill of $1 million. The roar of refrigeration fans fills the air: Below our feet, 6,675 tons of air-conditioning hardware labor to dissipate Dawn's body heat, blowing 2.7 million cubic feet of chilled air through the room every minute.

For a few days in 2009, Dawn brought its number-crunching muscle to bear on the largest brain simulation to date–a cell-by-cell model of the human visual cortex: 1.6 billion virtual neurons connected by 9 trillion synapses. This mathematical simulation of a cerebral cortex as large as a cat's blew away the previous record–55 million neurons, the size of a rat's–achieved by the same team two years before.

"This is a Hubble Telescope of the mind, a linear accelerator of the brain," says Dharmendra Modha, the project's lead researcher and a computer scientist from IBM's Almaden Research Center in San Jose, California. The simulation, assembled using neuroscience data from rats, cats, monkeys, and humans, will provide a tool for testing theories about how the brain works, and the insights gained could pave the way to designing robots and computers that are truly intelligent.

In his office at Almaden, Modha flashes an image on his computer. "This is an MRI of my brain," he says. Red and blue threads show the twisting routes of thousands of nerveaxon bundles that connect different parts of his cerebral cortex.

The cortex, the wrinkly outer layer of the brain, performs most of the higher functions that make humans human, from recognizing faces and speech to choreographing the dozens of muscle contractions involved in a perfect tennis serve. It does this using a universal neural circuit called a microcolumn, repeated over and over. Modha hopes the simulation will help scientists understand how the cortical microcolumn performs such a wide range of tasks.

But deciphering the microcolumn can also help build better computers. By reverse engineering the cortical structure, Modha says, researchers could give machines the ability to interpret biological senses such as sight, hearing, and touch. And artificial machine brains could intelligently process senses that don't currently exist in the natural world, such as radar and laser rangefinding. Modha envisions some day peppering the planet with sensors–transforming it into a virtual organism with the capacity to understand its own patterns of weather, climate, and ocean currents.

The simulation currently lacks the neural patterning that develops as real brains mature–complexity that comes from stumbling around in a body, in which every action has consequences. As Anil Seth, a neuroscientist at the University of Sussex in England, puts it, "The brain wires itself."

Seth and his colleagues demonstrated this principle, called embodied learning, at the Neurosciences Institute in San Diego, using a brain simulation called Darwin X. They embodied Darwin's 90,000 virtual neurons (roughly the brain of a pond snail) in a wheeled robot. As Darwin wandered around, its virtual neurons rewired their connections to produce the equivalent of hippocampal place cells, which help mammals navigate. Scientists don't know how to program such cells; with embodied learning, they emerge on their own.

Paul Maglio, a cognitive scientist at Almaden, has similar plans for Modha's cortical simulation. He's building a virtual world for it to inhabit using software from the video shootout game Unreal Tournament and topographic

THE *BLUE GENE/P SUPERCOMPUTER DEVOURS 1 MILLION WATTS OF ELECTRICITY THROUGH A RIVER OF CABLE.*

maps, aerial photos, and rover-level imagery from Mars. As the simulation moves around this virtual landscape in a rudimentary rover, programmed to avoid hazards that lead to injury, it will gradually learn the basics of eye-wheel coordination—just like a baby learning to walk.

The National Nuclear Security Administration (NNSA), within the Department of Energy, installed Dawn at Lawrence Livermore's Terascale Simulation Facility in 2009. After Modha's team ran its brain simulation, the supercomputer was transitioned to NNSA work, conducting massive simulations to ensure the readiness of the nation's nuclear weapons arsenal. Modha's neurons are far simpler than real ones. Yet for all its computing power, Dawn ran the 1.6 billion neurons at only one six-hundred-forty-third the speed of a living brain. A second simulation with 900 million neurons ran a little faster—but still at one eighty-third the speed.

HESE MASSIVE SIMULATIONS ARE MERELY steps toward Modha's ultimate goal: simulating the entire human cortex, about 25 billion neurons, at full speed. To do that, he'll need 1,000 times more computing power. At the rate that supercomputers have expanded over the past 20 years, that super-supercomputer could exist by 2019. "This is not just possible," Modha says, "it's inevitable."

But it won't be easy. "Business as usual won't get us there," says Michael McCoy, head of advanced simulation and computing at LLNL. Development of supercomputers in recent decades has ridden the wave of Moore's law: Transistors have shrunk and the computing power of processor chips has doubled every 18 months. But that ride is coming to an end. Transistors are now packed so densely on chips that the heat they generate can no longer be dissipated. To reduce heat, Dawn uses larger 180-nanometer transistors that were developed 10 years ago—rather than the 45-nanometer transistors used in desktop computers today. And for the same reason, Dawn runs these transistors at a sluggish 850 megahertz—one third the speed of today's desktop computers.

The supercomputer that Modha needs to simulate a whole cortex would also consume prohibitive amounts of power. "If you scale up current technology, this system might require between 100 megawatts and a gigawatt of power," says Horst Simon, a project collaborator at nearby Lawrence Berkeley National Laboratory. One gigawatt (1 billion watts) is the amount of power that the mad scientist Emmett "Doc" Brown needed to operate his DeLorean time machine in *Back to the Future*. Simon puts it more bluntly: "It would be a nuclear power plant," he says. The electricity alone would cost $1 billion per year.

The human brain, by comparison, functions on just 20 watts. Although supercomputer simulations are power hungry, Modha hopes the insights they provide will eventually pave the way to more elegant technology. With DARPA funding, he's working with a far-flung team at five universities and four IBM labs to create a new computer chip that can mimic the cortex using far less power than a computer. "I'll have it ready for you within the next decade," he says.

EDITOR'S NOTE

Building on its research in brain simulation, IBM recently announced the design of an experimental computer chip that operates according to the same principles as the human brain. Called "cognitive computing chips," these processors could lead to supercomputers that think like people. —DHW

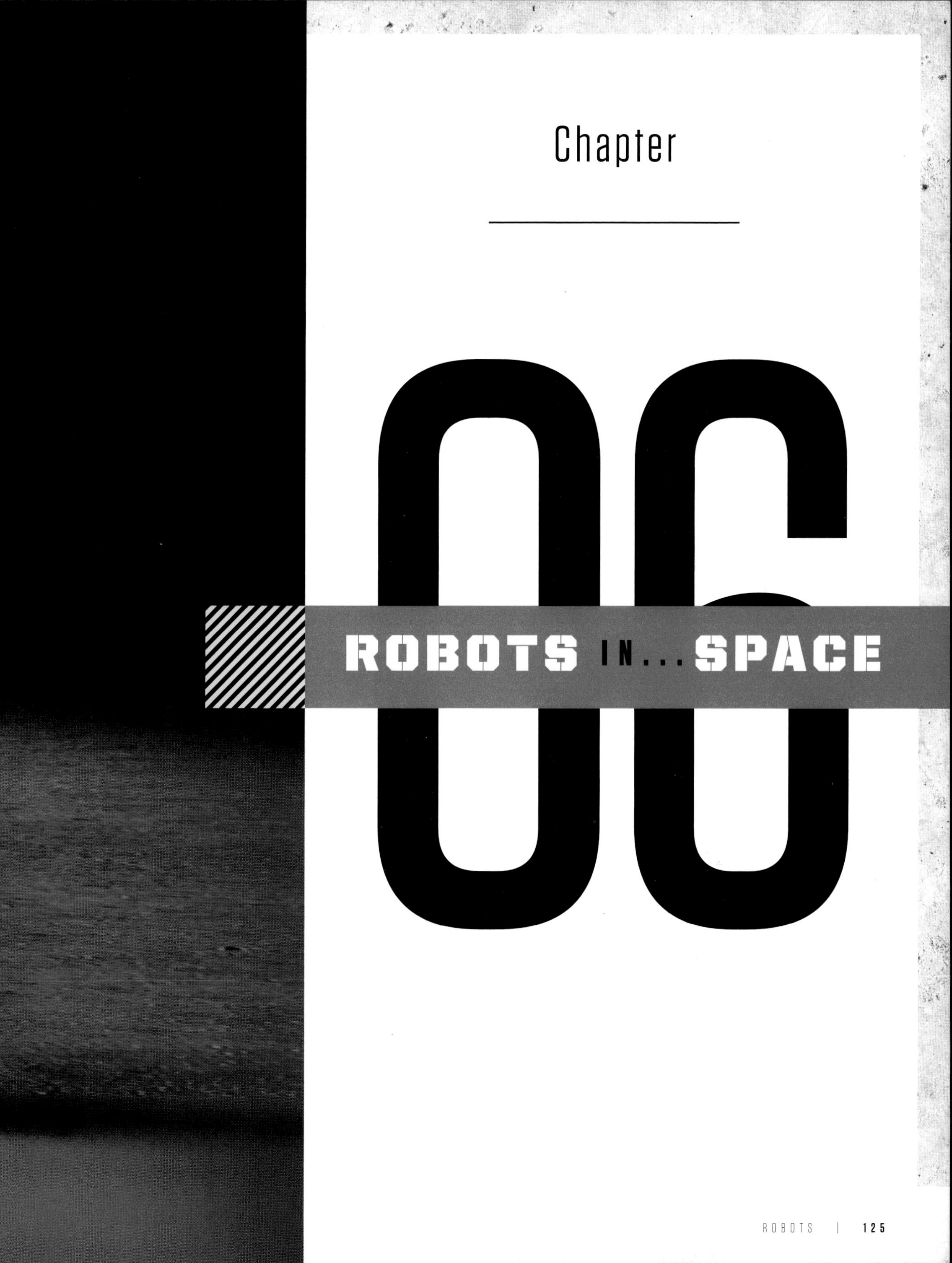

Chapter

06 ROBOTS IN... SPACE

HOW TO MINE AN ASTEROID

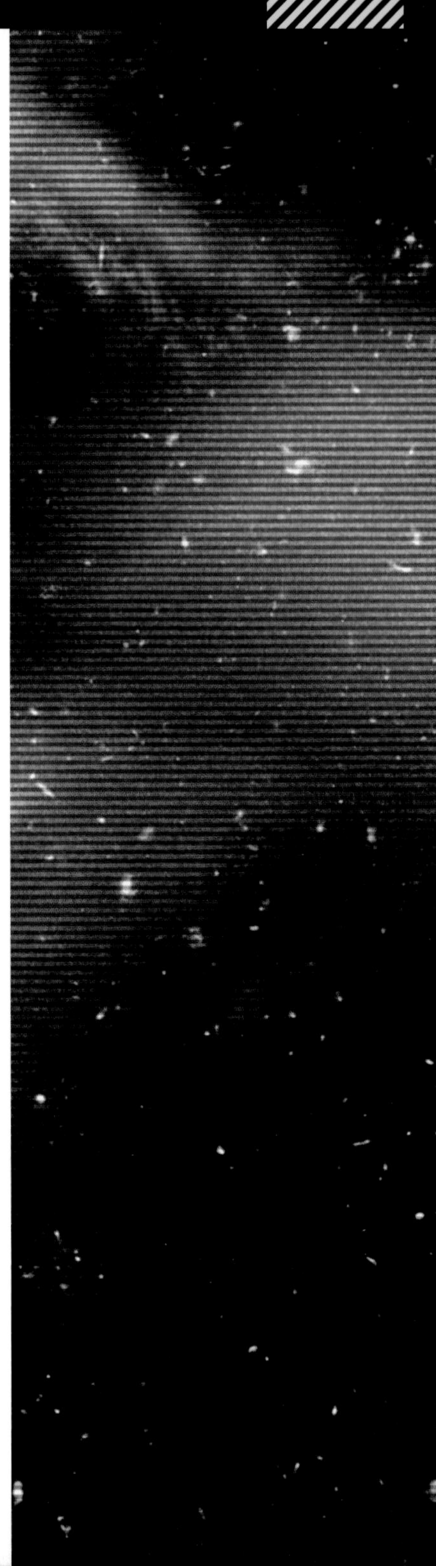

➜ In April, a group of aerospace veterans and investors announced an audacious venture: a company, Planetary Resources, dedicated to mining asteroids.

BY **MICHAEL BELFIORE**

This article was published in *Popular Mechanics* in 2012.

"**BREAKTHROUGHS REQUIRE TAKING EXTRAORDINARY** risks," says co-chairman Peter Diamandis. Planetary Resources, backed by technology trailblazers such as Google CEO Larry Page, movie director and inventor James Cameron, and Microsoft software guru Charles Simonyi, does not expect a fast return on investment. "Within a small number of years, we'll be flying to asteroids," says co-chairman Eric Anderson. "But we have a 100-year view for this industry."

STEP 1

GET PROSPECTING

TO MINE AN ASTEROID, PLANETARY RESOURCES first has to find one that promises a good return on investment. But asteroids don't glitter like stars. They are small, dark, and easily obscured by the distorting effect of Earth's atmosphere. The best way to hunt for them is with a telescope floating in space. At the Bellevue, Wash., headquarters of Planetary Resources, chief engineer and company president Chris Lewicki is assembling the components of the first privately owned space telescope, the Arkyd 100 series. The 44-pound spacecraft will be smaller and simpler than any government-funded space telescope. The $1.5 billion Hubble Space Telescope has a 94-inch diameter primary mirror; whereas Arkyd's mirrors will be 9 inches wide. Hubble has a wide field of view, as well as other instruments to scan objects in distant space. Arkyd needs only to look in our own solar system for targets. Being small saves money: Rockets carrying larger satellites could also haul these telescopes as secondary payloads, decreasing launch costs.

Planetary Resources plans to build a fleet of space telescopes to help drive the per-unit cost down to less than $10 million. Having multiple telescopes is insurance in case one fails. "We need to make something in an assembly line," says Lewicki, a former Jet Propulsion Laboratory Mars mission manager. "We can't just build one precious jewel that we treat with kid gloves."

The Planetary Resources team will also rent out the Arkyd 100s, the company's first stab at making money. Its space telescopes can be used by cosmic researchers or by Earth scientists who want to examine the planet from space at a resolution of about 6 feet per pixel. Planetary Resources hopes to launch the first satellite by the end of 2013. Company officials say rental prices have not yet been determined.

→ Space Mines Want Two Things

WATER

A 23-foot-diameter carbonaceous chondrite (C-type) asteroid can hold 24,000 gallons of water, which could be used to make rocket fuel or replenish spacefarers.

METALS

A 79-foot-wide metal (M-type) asteroid could hold 33,000 tons of extractable metals, including $50 million worth of platinum alone. But can a mining spacecraft cut off treasure from these metal objects?

STEP 2

ASSAY AND STAKE A CLAIM

ONCE COMPANY TELESCOPES SPOT A MINING prospect, there's only one way to determine what resources the asteroid contains: Get closeup.

The Planetary Resources team envisions a swarm of prospecting bots heading out to conduct close flybys of near-Earth asteroids (NEAs). "We're talking about building interplanetary probes at a fraction of the cost [of current models], which requires doing things very differently," Diamandis says. NASA has used this form of propulsion twice for deep-space exploration. It uses electricity to positively charge xenon atoms, which are pulled out of the craft by magnets. The repulsive force provides thrust that propels a vessel, building speed over the course of years. It takes a while, but when it gets going the craft can exceed 200,000 mph.

The asteroids of interest likely will be less than 1 mile in diameter, too small to have appreciable gravity. Spacecraft don't land on such small asteroids; they dock to them. A spacecraft will slowly approach, getting close enough to barely touch the asteroid's surface before deploying an anchor. Grappling hooks might just grab a chunk of surface material and float away. A better option is to deploy drills in each landing pad that secure the craft to the surface.

The robot would then analyze the water and metal content of the asteroid and beam the results to Earth. The tool of choice for this assay would be a laser-induced breakdown spectroscopy system (LIBS). Lasers vaporize surface material so sensors can analyze the light emitted by the resulting plasma to identify elements. The first LIBS to be deployed to another world, ChemCam, is currently en route to Mars aboard NASA's Curiosity rover.

The prospecting craft might also tag the asteroid by planting a radio beacon on its surface. According to company officials, the beacon would do more than help future missions get a fix on an asteroid's location. "Placing a beacon is part of building a case for ownership," Diamandis says.

A private company's claim to an asteroid is uncharted legal territory. In the next decade, lawyers may have to factor in the presence of private sector entrepreneurs in the Outer Space Treaty, first signed in 1967 and ratified by more than 100 nations. If it turns out that possession really is nine tenths of the law, then a simple radio transmitter could help make the miner's case.

One Asteroid To Go, Please

ASTEROIDS COULD BE BROUGHT CLOSER TO HOME FOR STUDY AND MINING. In an April 2012 publication, the Keck Institute for Space Studies, based at Caltech, looked at how to bring one to lunar orbit. Such a rock could provide an attractive destination for astronauts. "The mission will be a stepping stone into the solar system," says study co-leader Louis Friedman.

MEASURE IT

A slew of laser radar sensors measures the dimensions of the asteroid. A spacecraft then deploys its high-strength capture bag to the appropriate size. Inflatable arms and cinching cables unfurl to enclose the asteroid.

BAG IT

The spacecraft bags the rock. The finish on the bag's exterior ensures that the asteroid doesn't heat up and lose water.

BRING IT HOME

The craft makes the long trip back to lunar orbit. The return trip could take six years; mining commences on arrival.

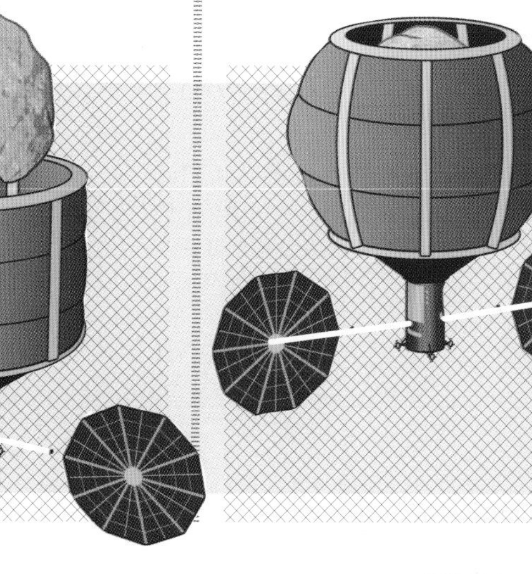

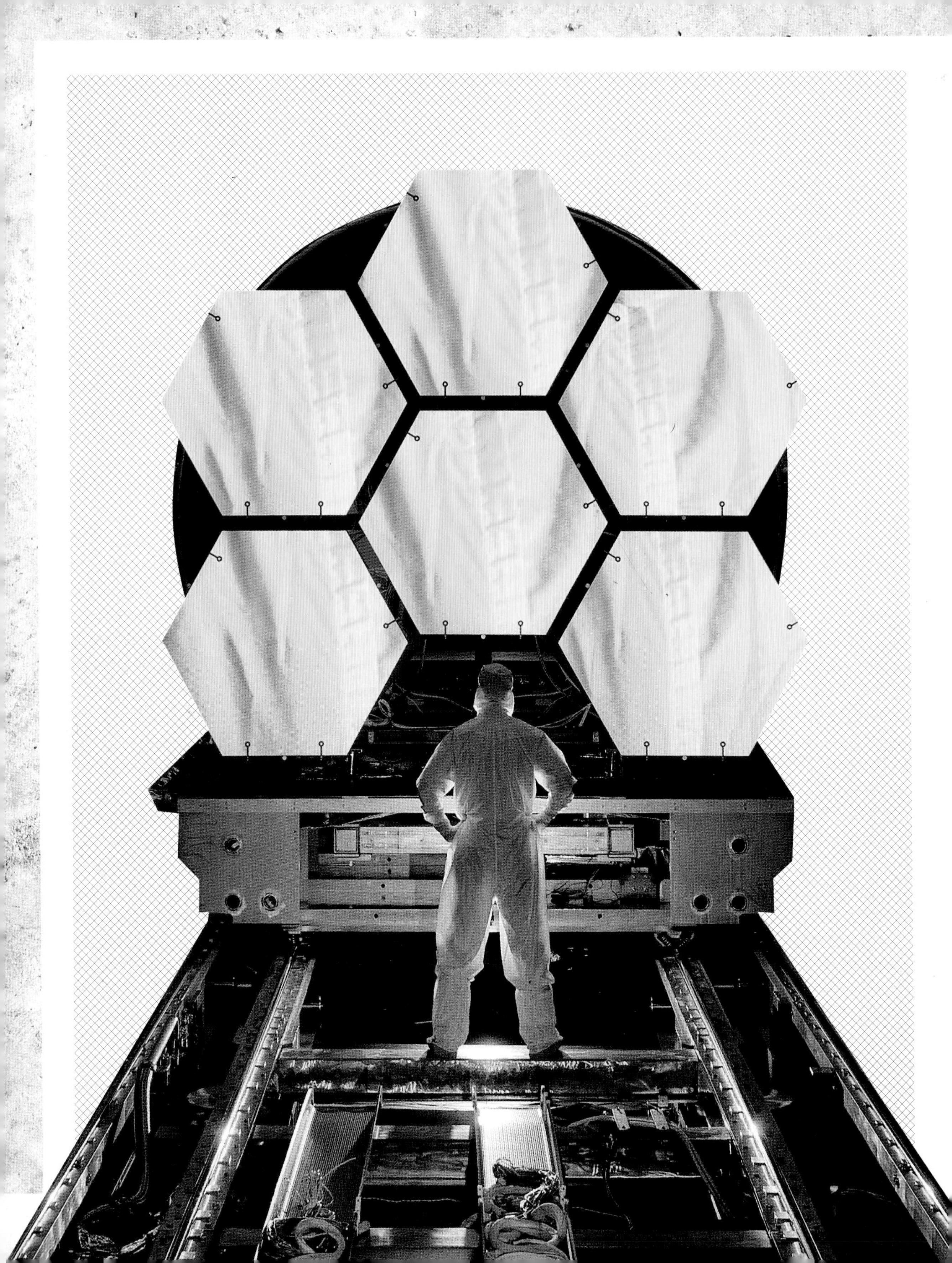

STEP 3

START DIGGING

SPACE MINERS WILL PRIZE WATER MORE than gold. Its value manifests when it is split into its elements: Hydrogen can recharge power cells and be recombined with oxygen to produce energy- rich fuel. Harvesting water in space is cheaper than shipping it from Earth. Every gallon, at a weight of 8.33 pounds, can cost tens of thousands of dollars to launch. Planetary Resources could profit by selling space-harvested water to governmental or private spacecraft at a premium but for less than it would cost to deliver from Earth.

Carbonaceous chondrite asteroids are the best prospects for water. The surface of these so-called C-type asteroids is crumbly, says John Lewis, professor emeritus at the University of Arizona and author of *Mining the Sky*, the seminal book on the space industry. "You can hold a cube between your thumb and your forefinger and crush it," he says. There's no need to burrow. You can just scrape the surface of a C-type asteroid to mine its water.

A swarm of mining bots, clinging with barbed feet to the surface of an asteroid, would slurp up water-laden soil through proboscis-like drills, while others would vacuum the debris left in their wakes. The robot would then pull out the soil, or regolith, and deposit it in a sealed container. The robot would walk, float, or crawl to a processor lashed to the surface or floating above it. The processor would heat the regolith to release water vapor, which would be collected into a storage tank.

Space miners face a more difficult challenge when harvesting metal. M-type asteroids, essentially big flying chunks of solid metal, might not feasibly be mined, says Harry McSween, geoscientist at the University of Tennessee and chair of the surface composition group for NASA's Dawn asteroid probe. Anchoring to such a body would be hard enough–drill-style landing pads wouldn't work, let alone sawing off a chunk of the asteroid for processing. "When you think about how much energy would be required, it seems pretty unrealistic," McSween says.

But Lewis figures that some asteroids might be made up of as much as 30 percent metal, in the form of an iron-nickel-cobalt and platinum-group alloy. "The temptation is to simply use a magnet to pluck the metal grains out of that regolith," he says.

Some metal-rich asteroids might be worth taking closer to Earth, as close as the moon, in their entirety. "The concentration of metal is so high that you have to wonder whether you could just bring the whole thing back," Lewis says.

STEP 4

DELIVER THE GOODS

SPACE SELLS, BUT WHO'S BUYING? It remains unclear who will purchase the goods that space miners have taken such pains to gather.

The most lucrative opportunity might be platinum-group metals–one category of the few space commodities that would be shipped back to Earth. "These materials enable so many different high-tech processes that we use," Lewicki says. Today, platinum-group metals are essential to catalytic converters in petroleum engines, as catalysts in the production of silicone, and in the manufacturing of glass. They are incorporated into hard drives; in spark plugs, where their low corrosion rates allow 100,000-mile life spans; and in medical devices, where they are prized for their biocompatibility.

A 500-ton asteroid with 0.0015 percent platinum metals–a common percentage–would have three times the richest concentration found on Earth. "To have more of this material will open up economies that we can't even predict," Lewicki says.

A NASA ENGINEER STANDS IN FRONT OF SIX SEGMENTS OF THE JAMES WEBB SPACE TELESCOPE'S PRIMARY MIRROR. SPACE MINERS MAY FIELD THE FIRST PRIVATELY OWNED SPACE TELESCOPES AND RENT THEM OUT.

→ Asteroid Mining Infrastructure

BETWEEN 2009 AND 2011, A NASA SPACE TELESCOPE CALLED THE WIDE-FIELD INFRARED SURVEY EXPLORER (WISE) CATALOGED ASTEROIDS IN OUR SOLAR SYSTEM. It found: more than 100,000 previously unknown asteroids in the belt between Mars and Jupiter, 19,500 midsize near-Earth asteroids, and 4,700 large, potentially hazardous asteroids within a cosmically close 5 million miles of Earth's path. NASA estimates that it has cataloged only 30 percent of them.

Orbital Transportation Hub

A larger, manned space station is an ideal place to coordinate flights of cargo, mining gear, and explorers.

Space Fuel Depot

Spacefarers will need places to restock water and hydrogen for fuel (think space gas stations). Scientists are working on ways to transfer fluids in zero-g.

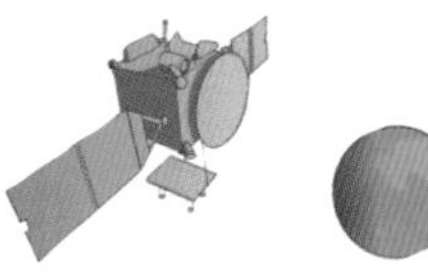

Deep-Space Communication Relay

Optical laser communications systems transmit as much data as radios and can use half the power. Planetary Resources is developing a system under contract with NASA.

Magnetic Rake

There is no need to dig mines to gather precious metals from space rocks. By placing a magnet on each prong of the rake, loose regolith (asteroid soil) can be combed easily for precious metals in low gravity.

Low-Gravity Sifters

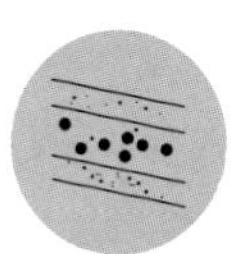

Heated asteroidal material releases oxygen. In 2009, scientists used a vibration table to shake soil through a sieve to separate the particles that would burn most efficiently in an oven. The system worked in zero-g, simulated by parabolic flights in an airplane.

Asteroid Anchors

With almost no gravity, asteroids won't be easy to land on, let alone allow for operating drills and other mining equipment. NASA's Jet Propulsion Laboratory is developing steel "toenails"; and Honeybee Robotics is creating screw-in augers to keep space machinery from floating away.

A *PROTOTYPE ROBOT, CREATED BY NASA'S JET PROPULSION LABORATORY, HAS 750 STEEL "TOENAIL" HOOKS FOR FEET. THESE ADHERE TO ROUGH SURFACES, PREVENTING BOTS WORKING IN LOW GRAVITY FROM DRIFTING AWAY.*

But most asteroid commodities will only be marketable in a future where ambitious spaceflight is a regular human activity; for example, extraterrestrial depots where spacefarers could top off their fuel tanks and water supplies while on long trips. If there are no such trips, there is no business model.

Similarly, the idea that common metals will be useful in space is predicated on a manufacturing industry that is building space stations and spacecraft in orbit. Assembling structures in space, rather than launching them from Earth, is appealing because it avoids the cost of launch. A lack of orbital construction or the advent of cheaper launch systems could obviate this business.

If space stations are growing food for full-time residents, they could become lucrative markets for more than iron and steel. Asteroid-derived nitrogen and ammonia would be in demand for fertilizer. Such industries are vital if humans are to make their home in space. "We're talking about technologies that break the umbilical cord to Earth," Lewis says.

Planetary Resources' scheme is more than a business plan. It's a rose-colored blueprint for supporting space exploration. Its existence speaks to humanity's drive to explore, to spread, and to support the most audacious of our dreams.

CURIOSITY *TIPS THE SCALES AT NEARLY A TON AND IS ALMOST 10 FEET LONG. PREVIOUS TWIN ROVERS SPIRIT AND OPPORTUNITY WERE 374 POUNDS EACH AND JUST OVER 5 FEET LONG. ALL THREE DWARF THE FIRST MOBILE MARS EXPLORER, SOJOURNER; IT WAS ONLY 23 POUNDS AND 2 FEET LONG. (ABOVE IS A FULL-SCALE MODEL OF CURIOSITY AT NASA'S JET PROPULSION LABORATORY ON OCT. 28, 2011.)*

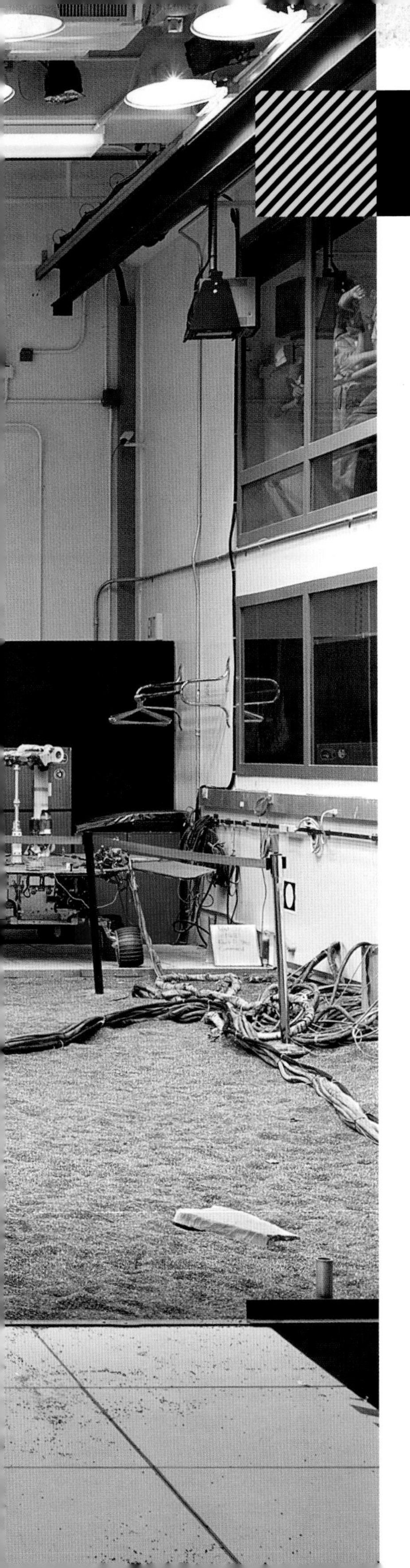

ROVERS

→ Robotic rovers constitute the vast majority of those who have boldly gone where no man has gone before—Mars. While exploring the red planet, each of these robots packs a unique approach, based on what secrets they seek to unveil.

BY **THE EDITORS OF POPULAR MECHANICS**

N AUGUST 5, 2012, AFTER NINE MONTHS in space, the 2,000-pound Curiosity rover made its descent to the surface of Mars via a so-crazy-it-just-might-work rocket-powered sky crane.

"Curiosity landed, drove 500 meters, and hit pay dirt," says principal project scientist and Caltech geologist John Grotzinger. That pay dirt was a gray soil sample dug up by Curiosity from just beneath

Mars' red surface. The soil's chemical makeup proved Mars once had enough fresh water to sustain life. "Modern Mars is red and inhospitable," Grotzinger says. "Ancient Mars was gray and habitable, if you were a simple microorganism." The Curiosity program has proved that it is possible to deliver heavy payloads to Mars and collect valuable samples. Next, NASA hopes to land a craft that can bring those samples back to Earth.

There's nothing better than a robot on Mars—unless it's one armed with a laser. ChemCam, a tool on Curiosity, can vaporize bits of rock from a distance and analyze the resultant plasma with three spectrometers to determine the target's chemical makeup. The tool helps researchers determine if the rover should use other instruments to learn more about the rock.

Curiosity's robotic arm drops rocks into CheMin's X-ray chamber; the way the rays scatter confirms the presence of clay deposits and sulfate salts. Liquid water creates these minerals. If conditions allowed water to exist on the surface for a long enough time, the chance that life existed on Mars increases dramatically.

If water is a key to life, then knowing how much is still there is vital. Curiosity has a particlebeam generator that can detect subsurface water ice. Shot from 2.6 feet off the ground, the beam cuts as deep as 6 feet beneath the surface. The way neutrons reflect to the surface determines how much ice is present.

Finally, SAM's gas chromatograph heats soil and rock samples until they vaporize and then separates the resulting gases into various components for analysis. This could produce direct proof of the presence of organic compounds trapped in rock—components essential to form life.

THE POWER SOURCE

Engineers hope Curiosity will rove for at least two years, relying on a unique power source. NASA has outfitted the vehicle with a nuclear-powered battery called a Multi-Mission Radioisotope Thermoelectric Generator (RTG). Drawing energy from the natural decay of plutonium-238 dioxide, an RTG produces almost three times the power of the solar panels that ran Spirit and Opportunity. And, unlike panels, RTGs perform well in dusty and dark conditions.

OR SIX-WHEELED, 400-POUND MECHANICAL geologists, Mars rovers Spirit and Opportunity have really turned on the charm, working their way into the hearts of an admiring public—and, most powerfully, those of their engineers. "We just couldn't be prouder of those little rovers," says NASA's deputy administrator, Lori Garver. After landing on Mars in 2004, each rover was meant to operate for 92 days and travel about three-quarters of a mile. Nearly seven years and 20 miles later, Opportunity still motors on. Spirit performed a feat of Martian mountaineering before finally running out of power on the darker, harsher side of the planet last year.

If the rovers' stamina wildly exceeded expectations, so have their discoveries: picture-postcard images from the surface of Mars, evidence that the planet once held water, footage of dust devils that offers insights into Martian wind. They also inspired the next generation of space explorers.

As John Callas wrote in a heartfelt goodbye to Spirit: "She has also given us a great intangible. Mars is no longer a strange, distant and unknown place. Mars is now our neighborhood." For all that, brave rovers, we salute you.

CURIOSITY *MARS SCIENCE LABORATORY: JOHN GROTZINGER (CALTECH, JET PROPULSION LABORATORY) AND JAMES ERICKSON (JPL) AT NASA'S JET PROPULSION LABORATORY, CALIFORNIA INSTITUTE OF TECHNOLOGY*

Anatomy of a Rover

WHEELS AND ACTUATORS

At 20 inches in diameter, Curiosity's wheels are larger than those of the average car–and twice the size of the sometimes balky wheels on the Spirit and Opportunity rovers. Each wheel features high-traction cleats and a dedicated motor. The front and rear wheels actually have two motors apiece, enabling independent steering. Impressively agile, Curiosity can turn 360 degrees in space.

THE SAM

Curiosity carries a 10-instrument package of tools and sensors– including the Sample Analysis at Mars (SAM) suite–to search for signs of life. One instrument features a remote-sensing device capable of targeting pinhead areas on rock formations from up to 23 feet away. ChemCam, as it is known, can then shoot a laser beam at the rock and analyze the vaporized material using light-reading spectrographs. "It has the potential to really study chemical evolution at a different scale than before," says co-investigator Diana Blaney. "We'll be able to map [material] composition at a fine scale." The SAM suite includes a quadrupole mass spectrometer and a gas chromatograph, which, along with the other devices, will work to analyze samples on the surface, inside Martian boulders, and below-ground, in the quest to find organic matter.

THE LANDING SYSTEM

Unlike its smaller roving cousins, which used airbags to cushion their touchdowns, Mars Science Laboratory (MSL) is far too large and heavy (nearly a ton) to land safely by bouncing. Engineers devised a unique, highly complex system called sky crane, essentially a rocket-powered backpack. In the early stages of descent, a parachute 165 feet in diameter deploys. The parachute jettisons and retro-rockets kick in, providing upward thrust to further slow the rover's descent. Radar detects the landing surface and orients the crane upright. Hovering with the aid of the rocket streams, the crane lowers the rover to the surface on cables, then releases Curiosity to begin its ambitious mission.

VOYAGER 1 & 2: NASA JET

NASA'S ***VOYAGER 1* SPACE PROBE IS POISED TO BECOME THE FIRST MAN-MADE OBJECT TO LEAVE THE SOLAR SYSTEM.**

LAUNCHED: ***1977***

SPEED: ***11 MILES PER SECOND***

DISTANCE: ***ABOUT 11 BILLION MILES FROM THE SUN***

EXIT: ***EXPECTED BETWEEN 2012 AND 2015***

PROPULSION LABORATORY

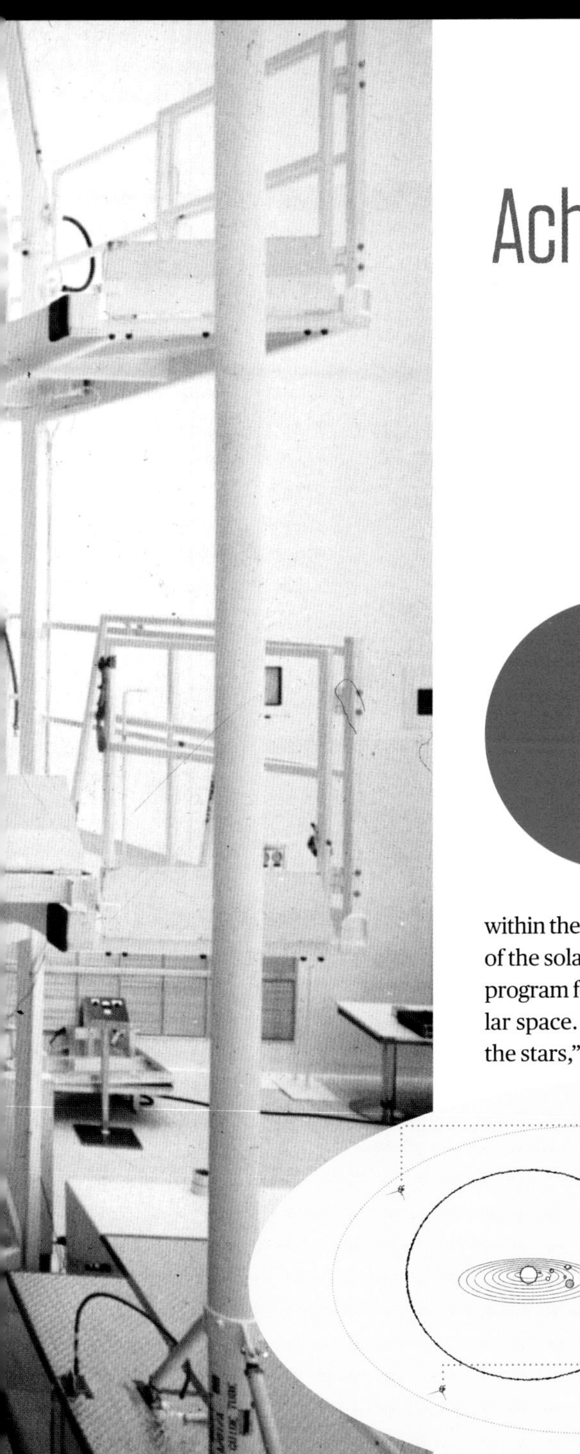

Mechanical Lifetime Achievement Award: Breaking Through the Heliopause

BY **THE EDITORS OF POPULAR MECHANICS**

IN 1977, NASA LAUNCHED TWIN VOYAGER spacecraft to take advantage of a rare alignment of the solar system's gas giants (Jupiter, Saturn, Uranus, and Neptune) that would allow both craft to swing past all four planets in a single trajectory. Engineers from NASA's Jet Propulsion Laboratory originally made plans for a 12-year mission. But by 1972, budget woes had withered their planetary grand tour to a five-year flight. Thirty-five years later, both probes are still sending back data, and within the next couple of years Voyager 1 and Voyager 2 will exit the farthest bounds of the solar system. Ed Stone, who has been JPL's project scientist for the Voyager program for four decades, is awed by the prospect of the probes entering interstellar space. "We have an object made by the human race that's traveling between the stars," he says. "It's not science fiction anymore. It's real."

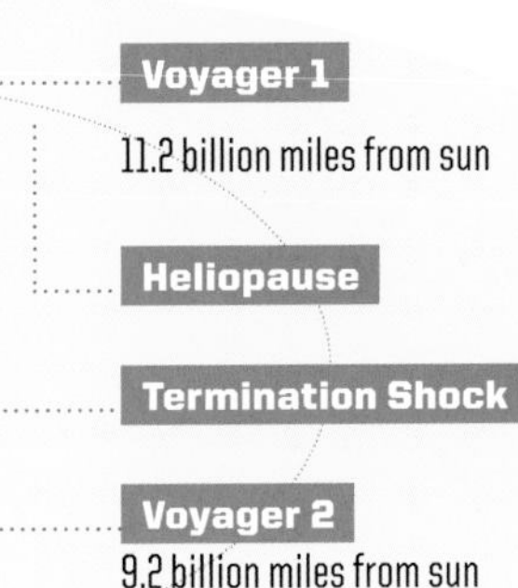

Reaching for the Stars
Both Voyagers have passed the termination shock, where the solar wind slows from supersonic to subsonic speed. The probes will soon cross the heliopause, the boundary where the solar wind essentially stops, held back by interstellar winds.

→ Go Up, Young Bot

ANDROIDS AND SPACECRAFT HAVE BEEN SCIENCE-FICTION STAPLES FOR DECADES, BUT FOR THE FIRST TIME THAT PAIRING IS BEING TESTED IN ORBIT.

The pioneering Robonaut 2 (aka R2) is aboard the International Space Station, preparing to go to work—and to validate the idea that a humanoid robot can be an asset to a busy crew. NASA and General Motors engineers designed the legless R2 to perform routine tasks—such as flipping switches, taking simple measurements, and polishing handrails—so scientists have more time for substantive work. "If Robonaut can provide just an hour's worth of relief to the crew, that would make the experiment worth it," says deputy project manager Nic Radford.

NOW

Getting Acclimated

Before R2 gets to work, NASA engineers calibrate its movements for a zero-g environment. "Now that you don't have gravity, there is no weight on the arm," Radford explains. "The robot moves with less effort." The adjustments ensure, for example, that the $2.5 million R2 won't punch a hole through an instrument panel while simply trying to push a button.

SOON

Learning on the Job

R2 will be given a task by engineers, then decide for itself how to complete it. "It's not an artificially intelligent robot, but it has the ability to do a task in an intelligent way," Radford says. Using visual processors and more than 350 force and torque sensors, R2 will be able to detect if it has correctly responded to a command, such as flipping a switch or tightening a bolt.

FUTURE

Stretching its Legs

On Earth, R2 weighs 330 pounds, so it's not a good idea to have it floating around a space station. NASA plans to send the lower limbs to the ISS to enable the robot to grasp handrails and swing like a monkey moving through trees. Robonaut 2 will also be able to dodge obstacles. "It'll say, 'Hey, something just floated by—I need to not run into it'," Radford says.

CADY *COLEMAN POSES WITH HER NEW SPACE STATION CREWMATE: THE FIRST ORBITAL ANDROID. FIVE CAMERAS, ONE INFRARED, ARE MOUNTED IN THE HELMET.*

INDEX

PHOTO CREDITS

Cover:
Front: © Gregg Segal
Back: © Anna Camille (Author),
© Noah Sheldon (top right),
© La Tigre (bottom right)

© Max Aguilera-Hellweg: 102
© Ian Allen: 91
AP/Wide World Photos: 18, 116 (lower right)
Corbis Images: © Jessica Hill/ /AP, 20
Courtesy Darpa: 45
Philip Friedman: 31
Getty Images: © SM/AIUEO, 36,
© Stan Honda, 52
© Harry Gould Harvey IV: 89
© Christopher Griffin: 58, 59
© Hibot: 66 (center)
© Mike Hill: 62
Chad Hunt: 84, 86
© Merce Iglesias: 51
© JPL–CalTech: 44 (right)
© Martin Laksman: 44 (left), 71, 90
© La Tigre: 108–109
© Tony Law: 23
© Joyce Lee: 32
© Spencer Lowell: 134
© Mark Mahaney: 17
© Dan Morison: 82
NASA: 138
© Brian Finke & Landon Nordeman:
4, 39, 41–43
© Pa c23: 128, 129, 131, 132
© Pokedstudio: 120
© Andres Rivera: 24, 25
Courtesy Romela: 53
© Brian Sarigul: 60, 61
© Gregg Segal: 8, 11, 14, 15, 19, 29
© Daniel Shea: 113
© Noah Sheldon: 100, 111, 112, 114, 115, 118
© Splitintoone: 13, 81 (top)
© Wei Leng Tay: 65
© Mark Peterson: 30
© Cody Pickens: 55
© Art Streiber: 26
© Vigilism: 48, 49
© Dan Winters: 80
© Nathaniel Wood: 79
© James Worrell: 47
Courtesy Wyss Institute: 57